LIFE IN THE
UNDERGROWTH

LIFE IN THE
UNDERGROWTH

David Attenborough

First published in 2005 by BBC Books,
A division of BBC Enterprises Ltd.,
Woodlands, 80 Wood Lane, London W12 0TT
www.bbc.co.uk

Frontispiece: a praying mantis in long grass, Europe

ISBN 0563 52208 9

Colour separations by Digitronix, Bradford
Printed and bound by the Bath Press, Glasgow and Bath

Contents

Foreword

We are greatly prejudiced by our size. We find it very difficult to believe that an animal that is many thousand times
smaller than ourselves can have anything in any way comparable
to our own motives, or to experience anything that resembles our
basic emotions of fear and hunger, let alone aggression or sexual
excitement. And until recently science abetted such thoughts.
Bees and blowflies, beetles and butterflies were mere automata,
mindless robots reacting automatically to the simplest stimuli. To
credit them with anything else was unjustified and scientifically
disreputable anthropomorphism.

I am reminded of an evening when we were filming one of the
television programmes that were being made at the same time that
this book was being written. We had in front of us a line of bot-
tles, each of which supported a spray of leafy twigs in which
crouched a small bolas spider. These tiny creatures catch moths by
whirling a filament of silk with a sticky blob at the end, whenever
one came near them. Kevin Fleay, the cameraman, had been
working with them for nearly a week and he introduced me to
them individually. This one, he told me, was very shy. The slightest vibration made her draw up her legs and stay motionless no
matter how near a moth came. That one reacted in the same way
if the light was too bright. A third didn't seem to mind how much
light was shone on her but on the other hand she was unpredict-
able. Sometimes she would hunt and sometimes not. But the one
at the end of the line, no matter how much she had eaten, or how
much light he shone on her, would whirl her bolas whenever a
moth came anywhere near and usually caught her prey. These

tiny little creatures half the size of my fingernail each had individual characters.

That experience alone reminded me of the dangers of making generalised statements about such animals. This book is an attempt to survey all the small creatures without backbones that live on land – in technical terms, the terrestrial invertebrates. When one considers that about a million of them have so far been identified by science and given names, and that there are probably twice as many still awaiting such recognition, that seems an impossible task. Almost any generalisation of any interest that one makes about a genus or a family, let alone a species, will have some exceptions. Nonetheless, such generalisations are worth making, providing their limitations are recognised, for only by doing so can this vast invertebrate world, which constitutes by far the greatest numbers of both species and individuals on earth, be comprehended, and only by tracing the broad history of its development, can the way the bodies of these animals are built and the manner in which they use them be properly understood.

Filming the behaviour of such small creatures in any detail has been very difficult indeed until quite recently. Even the most sensitive film stocks with the best available lens systems, required so much light that insects subject to it were in danger of being fried alive. Certainly they seldom behaved in normal ways under such conditions. These days, however, with the arrival of highly sensitive electronic cameras and the designs of new optical systems, such intense lighting is no longer necessary. Fibre optics now allow us to put lenses the size of pin-heads alongside our tiny subjects and we have ingenious mechanical systems that enable us to make those lenses travel alongside them as they move. The result has been to vastly improve our recordings of insect behaviour so that we can watch them behaving normally and in intimate detail and thus bring a new understanding and sympathy with the way life is lived in dimensions that are so greatly different from our own.

And that understanding has never been more necessary. The tiny invertebrate inhabitants of the undergrowth were the first creatures ever to colonise the land, over four hundred million years ago. While backboned creatures, including our own far distant ancestors, were still restricted to the seas, the invertebrates were making their way up the shores and into patches of low vegetation to establish the first eco-systems. And still today, their intricate inter-acting communities remain the very basis and foundation of life on land. Were they to disappear, the survival of bigger land animals that arrived later would be very difficult if not impossible.

It is my hope that this book and the television programmes that go with it will do something to improve our general understanding of these minuscule, beautiful, fascinating and essential creatures that preceded us on land and now make it possible for us to live here.

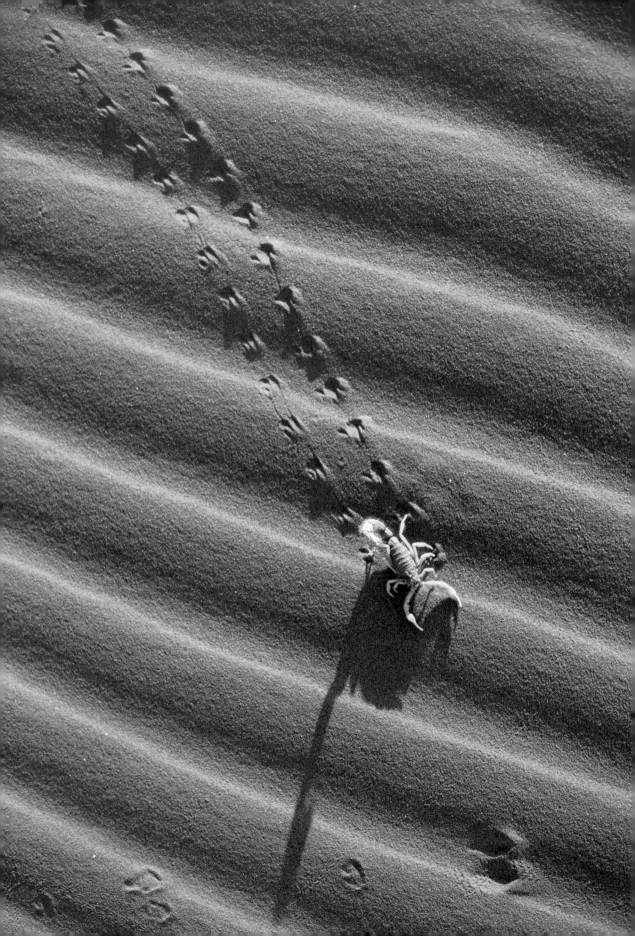

1

The Invasion of the Land

A spring evening on the coast of New England. A smooth sandy beach, separated from the flat land by a line of low grass-covered sand dunes, stretches for a mile or more to the distant horizon. The April moon is huge as it rises above the sea. It is full and as a consequence, the tide will be high. The wind is slight and scarcely ripples the silver surface of the sea. Conditions are good for what is about to happen.

A grey dome the size of a soldier's helmet appears in the white froth of the surf. It is so smooth that its wet surface glistens in the moonlight. As a wave swills up the shore and then drains away, the domed creature moving up through the retreating water becomes fully exposed. It has eyes, two small kidney-shaped swellings on the front of its otherwise featureless shell. They are slightly darker in colour and each has a surface covered with a thousand tiny facets. Attached to the back end of the dome there is a triangular section which ends in a rigid spike. This is jointed at its base so that it can be moved up and down and from side to side. Beneath, as you can see if one of the waves catches under the rim of a shell and flips the creature over, there are five pairs of jointed legs that slowly gyrate like parts of an overturned mechanical toy until, with luck, the animal gets a purchase with its tail spike on the sand beneath and manages to right itself, or another wave flips it back. It is a horseshoe crab.

As the moon rises higher, so dozens – then hundreds – and before long uncountable thousands of the crabs appear along the entire length of the beach. They have assembled on this particular

11

night to spawn. How they coordinated their movements so that they all arrived here at the same time is still not fully understood. Horseshoe crabs spend most of their lives deep in the sea, so in spite of the fact that they have compound eyes and even simple photo-receptors in their tail spikes, they live largely beyond the reach of light and they are unlikely to be able to use the shine of the moon as a cue. It may be that the speed of the tidal currents provides them with an indication of when to start their trek up to the beaches, for since a high tide, pulled by the moon's gravity, travels further up the beach than at other times, the water must then necessarily move faster.

A female as big as a large soup plate drags behind her a smaller male. He has attached himself to the back rim of her shell, holding on with a special lumpy claw shaped like a boxing glove at the end of each of his front pair of legs. She moves through the water in a surprisingly dainty way, just touching the sand with her legs, as though on tip-toe. As she advances into the shallows, her progress becomes more laborious, for she loses the support of water. When she reaches the farthest margin of the highest wave, she begins to tread with her legs, pushing away the sand beneath her so that she slowly sinks into it. And there she lays up to eight thousand eggs. Simultaneously, the male still clinging to her rear ejects his sperm. As she returns to the sea, so the wavelets wash sand into the shallow pit she made, burying the eggs. The eggs are sticky and about the size of the sand grains and will remain mingled with them for two to four weeks. Then the tide will once again wash up to the highest part of the beach and as it recedes, carry the developing larvae back to the sea.

The mass spawning continues for more than two hours, but as the tide turns so the crabs retreat and travel back to the deeps where they will remain for another year.

In some years, the tide is at its highest just before dawn. Then the last spawners are still ashore as the skies lighten. Now they are attacked by gulls. Those that have been overturned are now vulnerable. Gulls rip at their undersides and tear off their legs. But along the high tide mark, there are long black streaks – millions of

▷

The morning after a night of spawning, the New England shores are still thronged with horseshoe crabs. Gulls now come out to gather the crabs' eggs.

An overturned
horseshoe crab
reveals its five pairs
of legs, and behind
them the five pairs of
tightly closed plates
shielding the gills
through which it
breathes.

eggs that have been washed from the sand by the waves or maybe
thrown aside when one female crab started to dig where another
had laid eggs only a few minutes earlier.

There are also great flocks of wading birds – knots and dunlin,
sanderlings and turnstones. They are on their migration north to
nest in the Arctic, and the horseshoe crab beaches provide them
with an invaluable annual opportunity to refuel by feasting on the
crabs' eggs. Watching them probing the sand and collecting eggs
by the million, you can't help wondering why the crabs should go
to such lengths to deposit their eggs in a place where they are so
very vulnerable.

But the beaches of the world were not always so dangerous.
The horseshoe crab, which is only very distantly related to common shore crabs, is an extremely ancient animal. Fossils of a very
similar creature appear in rocks that were laid down five hundred
million years ago. At that remote time, the lands of the planet
were still totally bare and without life of any kind. No plants
clothed the rocks. No animals crawled or ran over the ground or
flew above it. The sea, on the other hand, was crowded.

Life had begun, as microscopic globules, some three thousand
million years earlier still. As the millennia had passed, increasingly

complex organisms had evolved and by the time the first horse-shoe crabs appeared, the sea already contained a multitude of different animals. On the sea floor, trilobites, like gigantic woodlice and ranging in length from less than an inch to more than a foot, trundled through the mud alongside sea urchins and starfish and sea snails. Above them, in the mid-water, cruised small jawless fish with bodies protected by an armour of heavy bony plates. Most formidable all, there were eurypterids, gigantic sea-scorpions some two metres long, with powerful grasping claws. These weapons and their great jaws make it clear that these creatures were hunters, preying on the smaller members of the marine community. In comparison with this sea, an empty shore must have been the safest of all places to deposit eggs. And it must still have its advantages in spite of all the hungry creatures that now live beside it, for horseshoe crabs have never abandoned the habit and each year they continue to re-enact what may well have been the very first invasion of the land by any animals of any kind.

A eurypterid or sea-scorpion, one of a group that dominated the seas over 400 million years ago.
▽

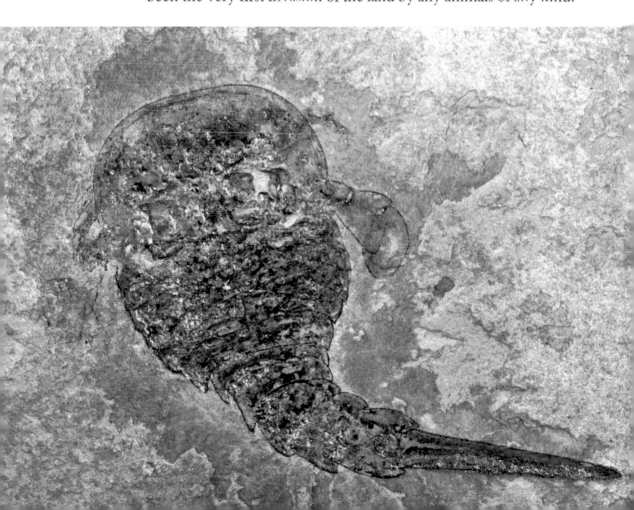

Horseshoe crabs can survive for some considerable time out of water. They normally find their way back to the sea by sensing the gradient of a beach and, after spawning, returning down-hill. But occasionally there is some confusion. Every now and then a crab may climb over the crest of the beach and then when the time comes to retreat, it goes down not to the sea but into the dunes at the head of the beach. If it finds a moist hollow there, it may survive for many hours, even days or weeks. It is almost as if it is pre-adapted for the terrestrial life.

A horseshoe crab breathes by means of five paired bundles of leaf-like plates under the rear half of its body. Wafted back and forth in the water, they absorb dissolved oxygen. On land they are capable of doing the same thing, provided they are kept moist. And a crab on land can also move around quite efficiently. Many water-living creatures, robbed of the support of water, are unable to make any progress at all on land. Jellyfish simply collapse. Most fish can do no more than flap from side to side and get no help from their delicate fins. But horseshoe crabs have shells – rigid external skeletons of a light, strong, versatile material called chitin – that enclose not only their body but also their limbs. Each leg is thus a jointed tubular rod and such a structure works very efficiently in or out of water.

But an external skeleton brings with it a major problem. Chitin bends but it does not stretch. It is not elastic. So a young horseshoe crab, if it is to grow, has to shed its shell and move into a new and bigger one. The process begins by a re-absorption of material from the underside of its armour so that its shell thins. Then a new, soft and crinkled skin forms under the armour, separated from it by a layer of fluid. The old outer armour begins to split and the animal, with wriggling movements of its soft flexible body, extracts itself from its old shell. It gulps water so that its body expands, smoothing out the folds of its new armour.

Slowly additional layers are added on the cuticle's underside that thicken and strengthen it. This process takes time and until it is complete the animal is very vulnerable, so such moults are usually conducted in some protected hiding place.

Growing in this way brings another limitation. If a large horseshoe crab were to shed its armour on land, its body would collapse under its own weight. In the sea, with the support of water, creatures with external skeletons such as lobsters and spider crabs can reach considerable size. The largest lobster ever recorded had a 50 centimetre (20 inches) body and weighed over 10 kilograms (23 pounds).

But that is minute compared to huge sharks and whales in the sea or lions and elephants on land. The skeleton that supports their bodies is of a fundamentally different design. It is internal – a backbone of vertebrae to which is attached a basket of ribs containing the internal organs and four sets of limb bones, the whole wrapped around with muscles and a tough skin. Bodies of that kind, vertebrate bodies, can grow uninterruptedly. But invertebrate animals with shelled bodies such as the horseshoe crab must moult – and that moulting must be counted as one of the factors that has prevented invertebrates from ever rivalling the size of the vertebrates on land.

Horseshoe crabs never moved permanently on to land but some descendants of their close relatives did. They came ashore not to lay their eggs but to feed. The remains of dead sea-creatures must have washed up on the beaches of the ancient seas just as they do today and they, it seems, were sufficient to enable some to feed on them and little else. And then other animals arrived that hunted the scavengers. Among them were the eurypterids.

A few of the terrestrial descendants of eurypterids survive to this day. There are a hundred and thirty species of them belonging to three different families. Most live in the tropics or sub-tropics and in those parts where the air is humid. You may find them under stones and logs, in crevices and on the walls of caves. Some have even taken to living in our homes, in dark corners, ventilator shafts or out-houses. One, the vinegaroon, is a midget compared

with its monstrous ancestors. It can be found in the warmer parts of the United States as well as South America and some parts of Asia. At the end of its two inch body, it has an equally long whip-like tail which gives it and the other members of its family the name of whip-scorpions. Its claws are strong enough to crush its prey and it defends itself by spraying acetic acid.

Another family, the amblypygids, are altogether more horrifying. Their bodies are scarcely bigger than the vinegaroon's, but their front limbs which sprout from either side of the mouth are greatly elongated and furnished with ferocious-looking hooks along the front edge and a sharp moveable spike at the end. These arms are almost twice as long as the animal's body and have a joint in the middle. The amblypygid is thus able to hold them out sideways and fold them back at the elbow, so that the spike at the end of each is in the front of its mouth.

Alarming though these arms are, they are not by any means the most unnerving of the amblypygid's hunting weaponry. Immediately behind them is the first pair of true legs. They are so long and thin that they seem more like antennae or tentacles than legs.

△
The vinegaroon, a descendant of the ancient sea-scorpions, that still survives in the southern parts of the United States.

▷
An amblypygid consuming a lizard which it grasps between its first pair of barbed limbs. Its second pair has been modified into thin mobile feelers with which it scans the rock face, one in front, the other behind.

18

They have internal thread–like muscles and so many joints that the animal can move them in any direction so they seem to writhe, more like the tentacles of an octopus than a beetle's antennae. In some species each of these antenna–like legs is about a foot long and with these the animal feels for prey. As it squats on a rock, one feeler sweeps round ahead of it, the other behind, gently tapping the rock.

They move with such delicacy that if they encounter an insect such as a moth, they may tap all over it without its victim even being aware that it has been detected. As soon as the amblypygid has fully located its prey, it shoots out its hooked arms to grab it. Usually it impales it with its terminal spike and then folds the spike back so that its prey is held firmly against a hook lower down on the arm. The prey is then brought back to its mouth where it is torn apart by the animal's blade–like mouthparts.

◇

There were also true scorpions in those ancient seas. They were similar in many ways to eurypterids and undoubtedly related to them. They did not achieve such size but they were nonetheless huge compared with their descendants living today. They too had a pair of pincers and excellent compound eyes, each with as many as a thousand individual facets. Their tail, however, did not end in a flat paddle as did that of many eurypterids, but with a spike.

Their breathing apparatus was slightly different from that of the horseshoe crab. Instead of books of thin leaves, they had beneath their abdomen absorbent membranes enclosed by a series of chitinous plates. These plates were attached to the abdomen by only one edge running transversely across the body so they could be flapped back and forth allowing water to reach the abdominal chambers above them where oxygen was extracted.

Scorpions too made the move on to land, though a little later than the eurypterids. Evidence that a particular species of fossilised scorpion was terrestrial comes from the shape of its mouth–parts. About 320 million years ago, species appeared with mouthparts that enclose a small space in front of the mouth. Land-living

▷
A desert scorpion in Arizona holds its prey with its pincers. Its tail is armed with a poison stinger and is held above its back but is used more for defence than for subduing prey.

scorpions today also have such a structure. When they catch their prey, they dismember it and then transfer the pieces into this little chamber. There they are soaked in a digestive fluid and dissolved so that the scorpion is able to suck in its food in liquid form. So the presence of such a structure implies a terrestrial existence. Drinking soup is hardly possible underwater.

Land–living scorpions also changed their breathing apparatus. The plates beneath their abdomen were reduced to only four and instead of being attached to the abdomen only along the top edge, they were fixed all round. Each plate is perforated by a pair of holes through which air filters into the chamber above. There oxygen is absorbed by a greatly folded membranous lining as it is in living scorpions today. They were almost certainly the first animals of any kind to become completely terrestrial.

In the millennia that followed, scorpions became smaller. By 300 million years ago, the largest was only about 30 centimetres (12 inches) long. Their eyes also became smaller and simpler and contain only about 40 facets. Microscopic examination of their exoskeleton shows that they also acquired a great number of sensory hairs that projected from their bodies. These two changes strongly suggest that the animals had become nocturnal hunters. Why should that be? By that time, the vertebrates – the first amphibians and reptiles – had also made the move to land. They were bigger and the scorpions, formidable though they were, lost their dominance and had to take cover beneath rocks and crevices during the day.

Today there are well over a thousand different species of scorpion and they can be found throughout the tropics and sub-tropics. The biggest of them, aptly called the imperial scorpion, lives in the humid rain forests of West Africa. Fully outstretched, it can measure 21 centimetres (8 inches). Some desert scorpions on the other hand are only a few millimetres long. All are very similar in form, with a pair of large and formidable pincers in front, eight legs, and a long segmented tail which is usually carried arched over the animal's back. At the end of it hangs a large tear-shaped stinger loaded with poison.

The potency of a scorpion's poison varies – as does the reaction of different animals when stung. As a general rule, the bigger and more formidable a scorpion's pincers, the less virulent its sting is likely to be. So the imperial relies more on its strength to overcome its prey and has a comparatively mild sting, whereas smaller species with fat tails and thin pincers have a venom so virulent that their sting can kill a dog in seven minutes and a human being in a few hours.

Although scorpions are normally only found in warm countries, their tolerance to different climactic conditions is extraordinary. They can withstand freezing for several weeks and will survive underwater for two days. Their external skeleton retains liquid so effectively that they can live in the hottest deserts. Their appetite is so small that individuals of some species can go without any food or water for twelve months. And they have a life span of up to thirty years.

If you try to pick up a scorpion, you quickly realise that it is almost impossible to catch the animal unawares. One way to make the attempt is to take a pair of forceps and grip it by its tail, just beneath the sting. But this is not easy. No matter from which direction you approach, the animal will be aware of you and will swivel to face danger often jerking its stinging tail forward in a threatening way. Almost as alarmingly, some species, such as a greenish black one that lives in southern India, will hiss at you, producing the noise by rasping a small patch of tiny spines on the basal segment of its claws against another patch of rounded tubercles on the inside of their first pair of legs.

They can, in fact, see you whichever way you approach for they have up to six pairs of simple eyes distributed around their carapace, together with a rather larger pair close to its back margin. So a scorpion can see both forwards and backwards simultaneously. These eyes consist of a group of individual light-sensitive cells. They have fewer elements than the large eyes of their marine ancestors and cannot therefore produce very detailed images. They can, however, detect the tiniest variation in brightness and so register movement.

As might be expected, direct sunlight might well dazzle if not permanently damage such delicate sense organs, but scorpions are able to deal with this problem by, in effect, putting on a pair of sunglasses. Each element in the eye mosaic contains granules of pigment which, in bright conditions, move upwards towards the surface of the eye, so forming a screen. When the light is reduced, the pigment drains away to the lower part of the cells. Scorpions are also able to see the ultra-violet part of the spectrum which is invisible to our own eyes. For some reason, the scorpions' external skeleton glows in ultra-violet light. It may be that this is just an incidental characteristic of scorpion chitin. It could, however, have some function. A scorpion's eyes are able to detect this fluorescence even at night under normal conditions, so it may be that the glow is of value to them in locating rivals or mates. It is certainly very useful indeed for scorpion scientists. They go looking for their subjects using ultra-violet lamps. In their light, a scorpion's armour shines with a surreal yellow-green glow so that the animal looks like some magnificently baroque enamelled jewel.

A scorpion in the Mohave Desert glows brilliantly in ultra-violet light.
▽

A few minutes searching will almost certainly reveal that there are far more scorpions in deserts than one might wish to know.

But vision is by no means a scorpion's primary sense. It will certainly be aware of your approach even if it cannot see you. The slightest movement on sand causes a minute vibration that is transmitted from grain to grain. The scorpion detects it with a slit-shaped organ on the upper part of each leg. These are so sensitive they can detect the footfall of a beetle on sand when it is a metre away. The difference between the stimulus received on one leg compared with that on another enables the scorpion to determine the direction from which the vibrations are coming. Air-borne noises are picked up in a different way – by minute hairs on the claws. With these it can detect the beat of an insect wing.

In addition to all these receptors the scorpion also has sensory devices that have virtually no parallel in any other animal – comb-like structures on the under-surface of its last pair of legs. They are called pectines and they are certainly sensitive for they

Pectines, the unique sense organs on a scorpion's underside.
▽

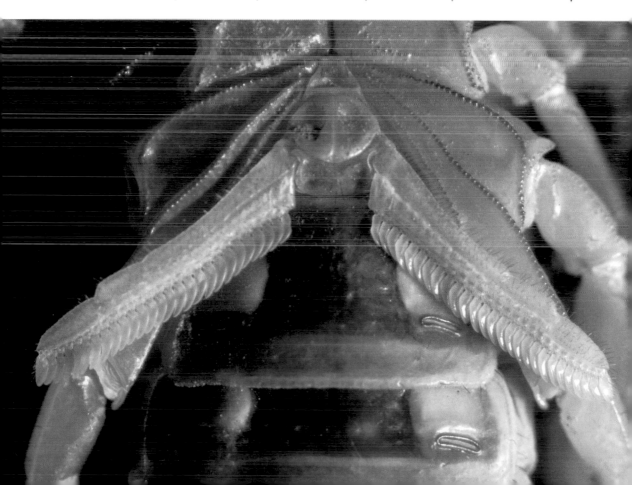

are packed with nerve endings. Until recently, in spite of all kinds of experiments by many investigators, it was still not certain what stimulus they register. Is it touch or hearing, balance or chemo-reception, a measure of humidity or the grain size of the sand beneath their feet? To complicate the question, in some species the males have larger pectines than females, so these organs probably have some part to play in sexual behaviour. It now seems certain, however, that they are chemo-receptors. Whatever their precise role turns out to be, part of it will involve the smelling or tasting of chemical substances on the surface over which they walk.

With so many different and sensitive ways of detecting what is going on around them, it is not surprising to learn that all scorpions are hunters. They prey on other small invertebrates such as cockroaches or even other scorpions of the same species, for they can be cannibals. Some will even tackle much bigger prey including snakes, lizards or small mice. But they seldom actively search for their prey. They normally wait until something accidentally wanders within their range. When it unluckily does so, they seize it with their pincers and transfer it to their mouth where they shred it with their sharp jaws. They seldom if ever use their stings in such attacks.

Plants were also making the move from water to land around the same time that animals did. Algae had been floating in the upper parts of the sea since the very beginning of life on earth and in due course they had become food for a multitude of different forms of marine life. Some in shallower waters had managed to attach themselves to the sea floor and become filaments or straps, but around 430 million years ago they moved out of water and on to the damp parts of the land. This probably did not happen on the shores of the seas for there the water moved back and forth, pulled by the gravity of the moon. But on the margins of lakes and the banks of rivers, in the permanent mists that surrounded waterfalls and in the bogs that developed where lakes became choked with sediment, there could be permanent moisture and there terrestrial algae were able to survive.

To advance further over the land and away from water, however, they had to develop a covering that was sufficiently impermeable to retain their internal liquids. In some this covering became rigid enough to enable them to stand upright. By doing that they were able to collect more of the sunshine they needed from ranks of simple leaflets on their stems. These were the first liverworts and mosses. They were followed by ferns and horsetails. So miniature forests developed that eventually provided more homes for creatures moving up from the water's edge.

Scorpions and amblypygids, like horseshoe crabs, have bodies that are divided into segments and so had many of the other creatures that also made their homes on the land – worms, millipedes, centipedes and the first, primitive wingless insects. This might suggest that all these groups sprang from the same stock. Although little fossil evidence has been discovered that might guide us in sketching the details of this remote genealogy, it is now generally accepted that this is indeed the case.

A segmented plan, however, was clearly a very effective basis on which to build bodies. Such organisms may have started with a single unit which possessed the basic equipment for an animal – a digestive chamber, excretory organs, a simple network of nerve cells, sex cells, and appendages of some kind to help the creature move, all enclosed within a protective skin. Such a simple creature, in the very early stages of life on earth, might have had the ability to reproduce itself by dividing into two and separating.

But if the units remained attached, then a bigger segmented body would be created. Nerve cells might join to form a continuous cord running from segment to segment along the length of the body, and digestive organs be linked to form a continuous gut.

The front segment clearly has special needs. It will require particularly good sense organs to inform the animal about what it encounters, so here the nerves become enlarged to service feelers on its head. Its digestive tract will have to remain open to the exterior as a mouth and its appendages will have to deal with the food it encounters. If more mouth parts are needed to deal with food, then several units at the head end may be merged to provide them. So the front section of the animal will become much more elaborate than the segments behind it. This process, if it happened, must have taken place at a very early stage in the history of life on earth – and in the sea.

The earliest marine communities of which we have any detailed knowledge date from some 540 million years ago. By this time, far distant though it was, segmentation was already well advanced, as exemplified by trilobites. Their hard shells are common in rocks of this period. They show that not only was the main part of their body divided into segments, each with a pair of legs on the underside, but that they also had elaborate head-shields equipped, in many species, with complex eyes.

But the soft parts of animals usually leave no trace, and there is no record of such animals that totally lacked shells – except in one or two very exceptional deposits. The first to be discovered were the Burgess Shales. They outcrop on the high flank of a peak in the Canadian Rocky Mountains. They were laid down as mud on the bottom of a shallow sea that sloped down very steeply from the nearby coast. Periodically, the mud slumped downwards in a marine avalanche, carrying with it all the small sea creatures that lived on it or in it. Their tiny bodies were immediately entombed before any marine currents could break them up or they disintegrated under the processes of chemical decay. Their delicate impress can still be detected, if the light strikes the shale at a particular angle, as glinting silhouettes. And among the trilobites was a remarkable soft-bodied worm-like creature that has been called *Aysheaia*.

Aysheaia had a long body with ten segments, each with a pair of stumpy legs. Each leg was itself ringed and carried a pair of minute curved claws that, in the right angle, catch the light particularly brightly. The front segment carried a pair of tentacles. Although other specimens of a similar age have been now discovered elsewhere, no others have been found later than those in the Burgess

One of the very first marine walkers, Ayshcaia, a 3½cm (1½ inch) 540 million years old fossil from the Burgess Shales.

Shales. Nonetheless we can be sure that the species or its descendants were still in existence at the time that the first miniature forests were becoming established on land – for some are still to be found in the undergrowth.

The first to be described scientifically was found perambulating through the leaf litter on the floor of the rain forest in Costa Rica. Its wondering discoverer called it, not unreasonably, a 'walking slug.' Its other colloquial name, 'velvet worm', is also apt for the animal's skin is covered with tiny closely-packed bumps of varying size, the bigger of which each carry a short filament, so that the animal does indeed have an undeniably velvety appearance. This first species was given the scientific name of *Peripatus* and that name is still used, inaccurately, for any member of the whole group. There are, in fact about a hundred and twenty different species and the family is technically called the Onychophora. That is a pity for that word means nothing more than 'claw-bearer', whereas 'peripatus' means, more evocatively, 'stroller' – not a bad name for an animal that was almost certainly one of the first creatures ever to actually live permanently and walk about on land.

Velvet worms have stumpy legs that are reminiscent of those of cartoon-like animals entertainers at children's parties make from artfully twisted tubular balloons. Many are brilliantly coloured – bright blue, red, and green, spotted and striped. They have two simple eyes at the base of fleshy antennae and their feet, like those of *Aysheaia*, carry a little double-hooked claw. They breathe through holes in their flanks.

They are all hunters. Their weapon is a milky liquid that instantaneously congeals on contact with the air. This is produced in two glands one on either side of the mouth and in such quantity that it constitutes about 10 percent of the animal's entire body weight. That is equivalent to a man carrying a weapon weighing about 5 kilos (around 11 pounds). A hunting velvet worm uses its long waving tentacles to locate its prey. As soon as it touches likely prey, it shakes its head and simultaneously squirts twin streams of this liquid, with the result that its victim is immediately entangled in a network of extremely sticky threads from which few can escape.

◁
A velvet worm from the tropical forests of Ecuador. The frills along its flanks are, in fact, its lobe-like legs.

A velvet worm attacks.
▽

Velvet worms are found in widely separated localities – in the Congo and Patagonia, the fringes of the Himalayas, tropical South America, Australia and New Zealand. Such a distribution is what one would expect from a creature that clambered on to land five hundred million years ago when many of today's continents were still joined together.

◇

The velvet worms seem to have only one group of close relatives among living animals. Very unexpectedly, these are almost microscopic. Most are less than half a millimetre long and are barely visible to the naked eye. Yet they are not merely specks of undifferentiated protoplasm like an amoeba with internally little more than a nucleus and one or two vacuoles. They have all the organs typical of advanced full-sized worms – a mouth with two miniature teeth, a stomach, a gut and a rectum, an eye-spot and a brain, sex glands and four pairs of legs, which like those of a velvet worm, are muscled balloon-like protrusions of the segmented body wall. You can find them almost anywhere on land that is moist – in gutters and ditches, on the bark of trees and bushes.

A water bear. Although scarcely bigger than a speck of dust, it has complex internal organs and eight legs.
▽

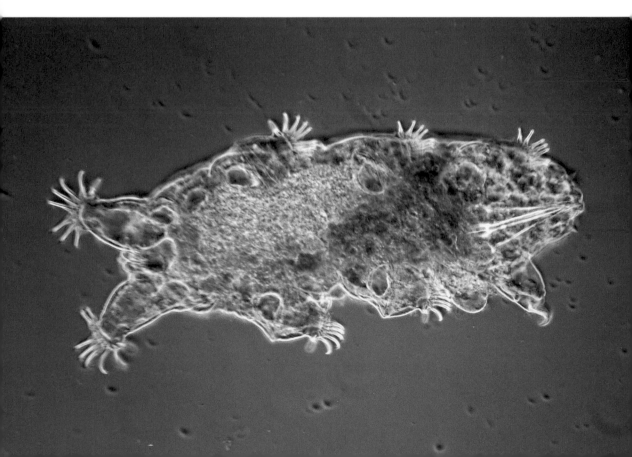

Their favourite habitat however, is in the water retained by damp moss. They are called tardigrades or water bears and there are about six hundred different species of them.

Under the microscope they have an engaging bumbling character as they use their stumpy legs to slowly waddle and barge their way through whatever minuscule fragments surround them. They feed by probing their tiny probosces into organic tissues, whether plant or animal, living or dead, and sucking out the liquid contents.

Although their skins are so thin that they are transparent, they regularly moult them and the females of some species, with commendable economy, use their shed cuticle as a receptacle for their eggs. Though water bears are most active in liquid, a mere film of water is sufficient for them. Only one or two live in freshwater or the sea. Should their precious water film disappear, they lose all the water from their bodies and simply shrivel. In this form, with virtually no liquid within them to burst their bodies should they freeze, they can withstand extreme cold. When better times return, they rapidly absorb water. Their tiny legs swell out and they set off again to find food.

There were other segmented invaders of the moist mossy forest as well as the Onycophora. The longest of them, with the most segments, are the millipedes. They do not have the book lungs of horseshoe crabs, but breathe through a series of holes in their flanks, a pair to each segment, that superficially resemble the pores in the sides of a velvet worm, although details of their structure make it clear that the two creatures developed these devices quite independently. These tubes run through their bodies and carry oxygen directly to their tissues. This system while highly effective in a small creature, becomes more and more inefficient with increasing dimensions and might also, like the problem of moulting, be a limit on size.

It may be, however, that the very first millipedes breathed with some kind of gill beneath the body in a way more akin to the

technique of their marine ancestors, for no trace of these flank pores has yet been found in the fossilised fragments of their bodies and they were giants, the biggest terrestrial invertebrates ever to have existed. They left tracks across the rippled sandstones of the ancient coal swamps some 300 million years ago, twin lines of imprints that are up to 36 centimetres (14 inches) across and look like the tracks of some motorised caterpillar tractor. The animals themselves had up to 30 segments and grew to a length of over 2 metres (6 feet). All millipedes are vegetarians and this one must have been an amiable monster – a kind of invertebrate cow – mooching its way slowly through the moist forest, munching moss and ferns.

Two millipedes. Above, a blue millipede from the forests of Mexico with flat armour plates on its back.

Below, a pill millipede from Borneo in its defensive posture. Both, like all millipedes, have two pairs of legs on each segment.

Though their name might suggest otherwise, no living millipede has a thousand legs. Each of their segments carries two pairs but even the biggest millipede today, a tropical giant from West Africa, has only about a hundred segments. Their legs project straight downwards from the underside of the body, and the animal, when it runs, lifts each pair rather laboriously in sequence, each fractionally later than the one immediately ahead of it, so that a wave of lifted legs seems to travel down the animal's long body. When alarmed, they do not try to run away but coil into spirals, like Catherine wheels. Those known as pill millipedes are much shorter. With fewer segments, they are able to roll up, head to tail, and so form a complete sphere with their helmeted head fitting neatly on to the armour surrounding the last segment. One species in Madagascar is the size of a golf ball.

◇

There were also segmented hunters – centipedes. They have far fewer legs than millipedes and they walk in a different way. Their legs project sideways from their body and are not moved simultaneously on each side, so that a running centipede wriggles and leaves behind it wavy tracks that are quite different from those made by millipedes. Some centipedes are very speedy indeed, and although they have fewer segments, many species grow longer than any living millipede.

All centipedes are carnivorous. Their mouths are equipped with mandibles with which they can cut up their prey, but flanking these they have a pair of huge fangs. These are in fact modified legs and are loaded with a most virulent poison. The colours of centipedes give a fair indication of just how venomous these creatures are for in the natural world bright colour is often a warning of danger. One centipede is a striking sapphire blue, another is emerald green with yellow bars across its body, and a third is mustard yellow with vivid blue legs. It has to be a very bold and reckless animal that interferes with such spectacularly coloured creatures as these.

They are very aggressive. Big ones will take a small bird or a mouse and will even rear up in the air to snatch a passing bee. A particularly alarming example haunts the caves of Venezuela. It is some 40 centimetres (nearly 16 inches) long, brown with each of its segments striped transversely in yellow and red. It is so muscular and so strong that those who have had the courage to grapple with one say that doing so is more like dealing with a small python than with a typical land invertebrate. This powerful creature hunts bats. It regularly climbs the rocky walls of caves and makes its way, upside down, across the ceiling to the bat roost. As it nears, it grips the rock with its hind legs and swings its body forward to strike the bat with its jaws. Such is the virulence of its poison that the bat is dead within thirty seconds. The centipede then feasts on the body as it holds it with its front legs, macerating the flesh with its mandibles for more than an hour. Local people say that a child was killed by a bite from one of these monsters.

It is usually assumed that centipedes like other small invertebrates rely on chance to lead them to their prey. The fact that these giant bat-killing centipedes make deliberate and regular journeys of four to five metres across rocky unproductive terrain, sometimes ignoring lesser prey such as cockroaches on their way, suggests that they are purposeful hunters. They can remember where prey is to be found and set out with a particular target already in mind. They must surely be the most formidable of all the hunters in the undergrowth.

▷
A giant centipede in a Venezuelan cave with a bat that it has just snatched from the air and is now devouring. It is over 35cm (14 inches) long and has only a single pair of legs on each of its segments.

The shortest of the early segmented creatures had only six legs. They make their first appearance in the fossil record some time after the millipedes and centipedes but still in a very ancient period – around 400 million years ago. The oldest fossils of them discovered so far come from cherts, a kind of flint, that occur in Scotland. Several groups of them still survive.

They are in fact, still widespread and numerous, though few of us ever see them. The most familiar, perhaps, is the silverfish for it invades so many of our homes. It is about a centimetre (a third of an inch) long with a shining white body covered in glinting scales and three long filaments at the end of its abdomen. It has a taste for anything containing starch and not only seeks out our larders but our shelves in order to nibble the dried glue and paste in the binding of our books. In the wild, they graze on lichens and algae as they must have done when they first appeared in the ancient forests. They have a close relative that also lives with us, the firebrat. It is similar in shape but prefers the comfort of well-heated rooms.

△
Silverfish have six legs and are classified as insects but, lacking wings, they represent a very early branch of that immense group.

△
A tiny springtail, dwarfed by the head of a pin. At rest, it keeps the tiny rod with which it is able to catapult itself into the air, tucked up beneath its abdomen.

Another kind, the bristletail, lives in the spray zone on rocky shores, nibbling sea weed. It too has three filaments at the end of its abdomen, though it differs from the silverfish in having the central one much longer than the other two.

Another surviving group, the springtails, are even more widespread. Indeed, they have claims to be among the most widely distributed animals on earth, for not only do they live on every continent, they also extend their range into both the Arctic and the Antarctic, where they occur in swarms on snowfields. They can also tolerate extremes of temperature. They live on the surface of tidal pools, in caves and in the fabric of birds' nests. Most are tiny – under 5mm (about ¼ inch) in length. They have a highly specialised two-pronged escape device on their underside. It is hinged to one of their segments and normally points forward. If they are alarmed, they flick it backwards with such power and speed that they are projected a foot or so into the air, so fully justifying their popular name.

Crustaceans, the predominantly marine group that contains lobsters, crabs and shrimps, also at some stage in their geological history made their own bid for a place on the land. Crabs started to feed on the detritus washed up on shores between the tide lines. Some of them advanced further still and dug burrows for themselves up on land where they could remain moist enough to breathe through soft absorbent membranes enclosed within a chamber of chitinous plates. The biggest of these, *Birgus*, the robber crab, is today the largest of all land invertebrates. The span of its legs is so wide it can embrace the trunk of a palm tree and its huge pincers so powerful that the crab can rip the husk from a coconut. But the land crabs' conquest of the land is not total. They have to return to water to breed. Some species do so in mass migrations, on the few nights of the year when the tides are at their highest. There in the surf the females shed their fertilised eggs. The larvae hatch almost immediately and remain growing and moulting in the sea for about a month. Eventually they

△
A robber crab. Its front legs can span a metre. Though it has to return to the sea to spawn, it is so well adapted to life on land, it would drown if it were to be underwater for any length of time.

acquire adult form and although they are still only the size of peas they then leave the water and take up residence ashore.

Relatives of the crabs, crustaceans known as isopods, include some species that have taken more radical steps. Like crabs, many are marine and others, known popularly as slaters, live on the seashore feeding on detritus. A few, however, have become totally terrestrial and are independent of open water, salt or fresh.

These are the woodlice, grey, flattened little creatures a centimetre or so long, that you can find in damp places, under stones or fallen logs, beneath bark or in the earth. They feed on almost anything – leaves both fresh and rotting, seedlings, fruit, decaying flesh, their own cast-off skins, even one another. Their segmented bodies are protected above by bands of articulated chitinous armour that stretch across their backs. Beneath, they have seven pairs of legs attached to their thorax. The rear section of their body, the abdomen, also carries appendages – plate-like gills. Each has two branches. The inner one carries a thin membrane through

A European woodlouse. It has excellent compound eyes and characteristic antennae with angles in them.
∨

which the animal absorbs oxygen. The outer one is more robust and helps to shield the inner one and prevent it from drying out. Some species have supplemented the gills with pads of tiny tubules close to the base of the gill through which they absorb air and aerate their blood directly.

A few species of these land-livers are able to roll themselves into perfect spheres like miniature armadillos. This may well have some value in defence, as it does for the true armadillo, for such a shape might well deter a spider or an ant. It might also prevent a larger animal with stronger jaws from recognising a rolled-up woodlouse as food. But the ability to roll up in this way most certainly helps in reducing moisture loss to an absolute minimum. Even with this help, however, this and most other woodlouse species restrict themselves to moist environments.

Life on land presents woodlice with one particular problem. Being crustaceans their eggs hatch into larvae. In the sea, these float in the surface waters together with other members of the plankton. How are they to develop on land? When the breeding season approaches, the females develop plates at the base of their front five pairs of legs. These grow until they meet in the centre of the underside of the thorax and so enclose a small chamber which the female keeps filled with water. She lays her eggs directly into this pouch and here they are fertilised by the male.

Woodlice, like all creatures with exoskeletons, have to moult, but have their own unique technique, shedding their old skeleton in two halves, front and back.

After several weeks of embryonic development, the eggs hatch into tiny larvae. At first they are incapable of movement but they develop rapidly and soon they are, in effect, swimming within the pouch. In some species there may be as many as 200 of them. Within a few days they emerge and after twenty-four hours in the soil they undergo their first moult, taking their first step towards adulthood and a life on land.

Some of the inhabitants of the ancient seas were able to permanently colonise the land with hardly any anatomical changes, for now it was possible to find shelter among the litter of dead leaves and within the soil that began to accumulate on the ground beneath the tall tree-like ferns and horsetails.

Segmented worms, not unlike the bristle worms that can be found today beneath rocks along the sea shore, also inhabited the sea floor alongside *Aysheaia* in the Burgess Shales seas. Their descendants eventually wriggled ashore and burrowed into the ground to live, presumably, in very much the same way as earthworms do today.

The bodies of living earthworms are enclosed by an extremely flexible cuticle. Consequently they are able to shorten or extend themselves by muscular contraction. To advance, they fasten the bristles in the back segments of their body into the surrounding soil and then contract the muscles surrounding their front section so that their pointed tip pushes into the soil ahead. The front section then expands and takes a grip on the soil, while their rear end slims and is pulled forward up the burrow.

They feed by ingesting soil with the mouth at the front tip of the body. This passes down into a muscular section of the gut, a kind of gizzard, where it is ground into a fine mud, to extract such vegetable detritus as it contains and then voided from the back end. Those two actions – burrowing and feeding – alter the nature of the ground in which they live, aerating it and breaking down its particles so creating a rich loamy soil. Worms therefore were very important factors in the establishment of the early forests. And they remain so today, essential elements in the fertility of the land, the very basis of so many ecosystems. Where they thrive, they may number seven million individuals in a hectare and those in an English meadow can, together, weigh more than the cattle that might be grazed there.

Worldwide, there are today over three thousand different species of earthworm. Here and there, in isolated pockets in New Guinea and South Africa, Sri Lanka and South America there are giants. It seems that a family of worms in each area separately evolved a giant form. The biggest of them is a species that lives in Australia. Its range is tiny, a small patch of what is now agricultural farmland a mere thirty kilometres square in Gippsland in the state of Victoria.

The exact length of these Gippsland giants is difficult to establish for they are, by their very nature, extensible. If one is freshly dug up, it may measure about a metre. But if it is then held up by one end, a practice often adopted by those investigators hoping to establish records, the weight of its body may cause it to stretch to about twice that length. That treatment, however, may seriously damage the worm. Its muscles are not strong enough to support such a great weight and quickly become exhausted. Its body then develops a number of constrictions so that it resembles a string of sausages and the worm may break into separate fragments and die.

Little is known about the natural history of giant earthworms, for they seldom if ever come to the surface. You can, however, hear them. As you walk through the pleasant pastures of Gippsland, particularly in gullies and near streams where the land is well watered and the soil moist, you may suddenly – and very surprisingly for those not prepared for it – hear the sound of water gurgling down a plug-hole. It is a giant earthworm squelching along its waterlogged tunnel in the soil a few inches below your feet. These tunnels are scarcely wider than a human finger for the giants, while very long indeed, are nonetheless quite thin.

The worms create a permanent network of inter-connected galleries that extend downwards for between one and two metres to where the soil is permanently moist. They spend their entire lives underground stolidly eating the earth ahead of them, extracting the vegetable material it contains and ejecting it from their back end. Most species of earthworm feed this way but then wriggle up to the surface to eject this processed material as worm-casts. The Gippsland Giant however, packs it away at the end of the chamber it currently occupies. How the worm creates the space for its extensive galleries is not properly understood. It must presumably be produced by compacting the soil around its tunnels by flexing the muscles of its body.

Every year or so, an adult worm deposits a single egg cocoon in a side-chamber of one of the galleries. It is about six centimetres long and looks remarkably like a cocktail sausage. Its thin brown chitinous skin is semi-transparent so that if you hold it up to the light, you can see a single infant worm within, slowly writhing in fluid. Such cocoons, taken to a laboratory may take up to a year to hatch. The juvenile that emerges is about 20 centimetres (8 inches) long – already considerably larger that most European earthworms. But no one has ever found more than one adult worm in a tunnel system and no one has any idea how and when two individual giants might meet to achieve cross-fertilisation.

Snails also existed in the ancient seas and must have lived in very much the same way as their living descendants today. They moved around on a flat fleshy foot, fed by rasping vegetation from the surface of rocks with a file-shaped tongue and breathed by sucking oxygen laden water into an internal pouch. All those techniques work very well on land. A pouch in the flank can absorb oxygen from the air just as it can from water providing it is kept moist. With glands beside the mouth producing abundant slime, snails can move around with as much ease as they do in the sea.

The only extra feature they need for a terrestrial life is a method of sealing the entrance to their shells to prevent themselves losing moisture when circumstances become dry. Some produce a small disc of chitin with a chalky top layer that is attached to the side of their bodies and fits neatly into the mouth of their shell when they retract their bodies into it. Others use mucus to plug the entrance to their shells or cement themselves to a stone or the branch of a tree where they will stay until their surroundings become moist again.

In the sea, snails such as the conches can grow to 60 centimetres in length and 2.5 kilos (5 pounds) in weight. Those living on land, without the support of water, cannot match that. The largest, *Achatina*, the giant West African snail, reaches a length of 30 centimetres (12 inches). Doubtless the weight of a shell becomes a considerable burden above a certain size. It would also be a serious encumbrance for any creature that tried to spend much of its time underground and some small snails eventually discarded it.

These are the slugs. The only sign that their ancestors once owned shells are a few chalky granules or a tiny plate, hidden within the flesh of their back. Unprotected, they have become nocturnal, only sallying out to feed under the cover of darkness when few predators are abroad. During the day they are very skilled indeed in finding cracks and crevices, under stones and beneath bark, in which to hide. Many a gardener is in happy ignorance of the extraordinary large numbers of these creatures that browse in his garden at night. One zoologist removed between ten and seventeen thousand slugs a year from his suburban garden for four years without making a significant reduction in their

△
Originally from Africa, this giant snail has now spread to India and the Far East. It can weigh over 200 grams (7 ounces)

numbers – a testament not only to the size of the slug population but their efficiency in reproducing themselves.

So even before back-boned animals arrived on land, the swards of moss and liverworts, and the ranks of horsetails and tree ferns had been colonised by a wide variety of creatures, all of which had ancestrally come from the sea. But though they had all, in their different ways, solved the problems of moving around and breathing in air, they all also faced one other major problem. Mating.

◇

In the sea there had been little difficulty. The simplest solution, adopted by corals, sea urchins, some worms and many other kinds of animal, is simply to squirt eggs and sperm into the water and allow them to mix. Horseshoe crabs also use this system even though the male marginally increases its efficiency by clinging to the back of a female and releasing his sperm as she sheds her eggs. But the technique is obviously very wasteful and even in the sea the males of a number of species, among them shrimps and shore crabs, package their sperm in bundles, known as spermatophores, and deliver them individually to the female.

Out of water, the need for personal and intimate delivery is even greater than it was in the sea and the variety of different ways that these early land-livers have developed in order to do so is quite astonishing.

The velvet worms use what is perhaps the least complicated system. It is of an almost alarming simplicity. The male simply sticks his sperm bundle on to a female's body. He does not seem to be very particular about where he places it. He might put it on her flank, on her back or on her head. Nor is he very discriminating about the identity of the individual to whom he attaches it. Another male or a juvenile is quite likely to receive one if they come his way. If it lands on a female however, it seems to melt into her. The tissues beneath it dissolve. The velvet worm's internal anatomy is so simple that it does not have a system of arteries and veins. Instead, its internal space is filled with a general all-purpose liquid and the sperm spreads through this until some of it

happens to reach an ovary. The young then develop internally and do not emerge from their parent's body until they are fully developed and able to assume independent life.

Earthworms are hermaphrodites. Each individual possesses both ovaries and testes. But this does not reduce the problem. Cross fertilisation is as necessary for an earthworm as it is for other creatures where the sexes are separate. To describe the preliminaries to their mating as courtship is, perhaps, to strain that word somewhat. Nonetheless, there are certain preliminaries that have to be performed by worms before they copulate.

On a warm spring evening, when the grass is wet with rain or dew, earthworms living beneath an English lawn venture out of the safety of their burrows. These usually have two entrances, one

△

A velvet worm gives birth. The young is born ready for independent life with antennae already fully developed.

which descends more or less vertically and another which is inclined at a more gentle angle and joins the vertical one an inch or so below the surface. It is from this angled burrow that the worms emerge and wriggle through the damp grass, cautiously exploring their surroundings while their rear end remains in the burrow, keeping a firm hold on it. Should a worm be disturbed by a bright light, a firm foot-fall or some other vibration in the ground, it will immediately and swiftly contract and withdraw into its burrow for safety. But if it is undisturbed, it may well encounter a neighbour, similarly engaged in exploration. Sometimes the contact seems to elicit little reaction. But on other occasions, the two will linger together. One will then extend itself still further and squeeze its front end into its neighbour's tunnel, while still maintaining a grip on its own burrow with its tail. Sometimes the neighbour retreats as the other probes the burrow, sometimes the questing one has to squeeze past the other's head end. Then both withdraw. The visited now becomes the visitor and stretches across to its neighbour's hole to make its own investigation.

There may be a dozen or so exchanges of this kind during the night or over one or two others following, but eventually both partners seem satisfied with one another, for both emerge at the same time and meet midway between their two burrows Their front sections overlap and they hook fine bristles on their flanks on to one another so that they are drawn into a close embrace. Each animal has a yellowish band, a girdle that encircles several segments of its body. As they lie there, mucus exudes from these girdles and covers over the overlapping bodies, enclosing them both.

Within, one of the partners ejects sperm from its testes which, in the commonest of English species is on the fifteenth segment from the head. Very slowly, so that the movement can only be clearly seen in speeded-up film, the two bodies pulsate. Sperm is travelling down a groove between the two closely clamped bodies and into a pouch, a kind of holding bag, of the other. While this is happening, it is not unusual for a third individual from a nearby

burrow to move across and start investigating the paired couple, apparently imbibing some of the mucus they produced.

The paired worms don't react to this. They may lie together for up to three hours during which time sperm passes along the passage between them in both directions. And then, at last, they separate and retreat, each to its own burrow.

Several days later, each secretes more mucus from its girdle. This hardens to form a broad ring which the worm pushes up its body with a series of spasmodic muscular jerks. As the ring passes towards the head it collects first its partner's sperm from its holding pouch and then eggs from its own ovaries in a segment a little further up. At last the ring with its enclosures reaches the animal's head. The worm pulls itself free, and the open ends of the ring contract to form a neat oval capsule. This then lies in the earth until in due time the young worms within wriggle their way out of it.

△
Earthworms busy exchanging sperm on a damp summer's evening.

A pair of small garden slugs investigate one another with their antennae while exchanging sperm.

Slugs are also hermaphrodites but for them, the exchange of sperm is a much longer and more complex process. Each individual has two pairs of tentacles on its head. The longer pair carries an eye on each tip. The shorter is sensitive to chemical perfume and it is this pair that tells a slug seeking to mate that the slimy trail it has encountered was left by another slug with the same disposition. It sets off in pursuit waving its tentacles excitedly. The first slug now begins to crawl in a circle. The other follows doggedly. They chase one another for maybe an hour or so with the size of the circle steadily diminishing until at last the two are moving close to one another, head to tail and flank to flank. Each extrudes its penis and they exchange sperm.

The leopard slug's courtship is more complex. The pursued one habitually heads for an overhanging surface such as the branch of a tree or a bracket fungus, one of their favourite foods. They both climb up and resume their circling there.

Suddenly one raises its front end and crosses the body of the other. The pair, entwined, now curve away from the overhang and start a slow-motion dive, twisting around on a double rope of slime, hanging on to it with the tips of their tails. When they have descended some 30 centimetres (12 inches), they halt their slide and each extrudes its penis, which entwine. As the pair tighten their coils around one another their penises suddenly flare outwards like an umbrella and they exchange sperm. Then they separate, each carrying the sperm from the other. One of them reaches out and touches the nearby rock or tree trunk and crawls on to it. The other, now by itself, climbs back up the mucus rope.

The banana slug, which lives in the forests of the western United States, is even larger than the European garden giant, reaching lengths of 25 centimetres (10 inches). It behaves in an even more extraordinary way. A pair, after beginning their circling courtship dance, strike at one another's flanks with their mouths, inflicting large wounds on one another. They lift their fore-parts and slam down on to the ground. They evert their penises from their flanks and wave them at one another. This performance may continue for twelve hours before at last they exchange sperm. Having done that, one partner often bites off the other's penis and eats it.

Millipedes, moving on their multitude of legs instead of sliding along a carpet of mucus, do not leave smelly trails behind them, so an individual has to have other ways of attracting a mate. Their sexes are separate and a male, ready to breed, uses sound to broadcast the fact, in some species by rasping one of his legs against his body. The female responds by coiling up in a tight spiral. It is probably her way of selecting a mate, for any male that wins her has to be strong enough to force her coils apart. Once he succeeds in doing so he coils himself around her and begins to writhe so that he moves up her body in a spiral.

The sex organs on both female and male are placed on the second segment behind the head. But the male lacks a penis. Instead he has a leg on his seventh segment that has been specially

▷
A pair of leopard slugs, in the middle of their acrobatic courtship.

modified to serve as an instrument of sperm transport. Before he achieved contact with the female he prepared himself by reaching forward with this sex leg and extracting a small sperm bundle from a cavity on his second segment. Now he uses that leg to carefully insert his sperm into her sexual pouch.

In some species the male's seventh leg has a particularly elaborate structure. It is armed with spines and ends in a scoop. A female can retain sperm in her pouch for up to six months and will mate with several different males if she has the chance. So a male having achieved contact, first uses his sex leg to scoop out any sperm that may have been deposited there during an earlier mating. Then having replaced it with his own, he will remain entwined for up to two hours to give his own sperm some chance of fertilising an egg before he has to move away and leave his sperm at the risk of removal by a successor.

The male of some species of bristletail has a particularly

△
Courting millipedes coil around one another, head to head. Here the male clasps the female's body with his pink legs and uses a specially modified one to transfer his sperm to her.

ingenious way of offering his sperm to a female. He too begins with a dance. From the way she responds to his first steps he is able to assess her readiness to mate. The pair stimulate one another with their antennae. Then he produces a thread of silk from his rear end. As it emerges, he attaches three or four drops of sperm on it. He sticks one end to the ground and then lifts his abdomen so that the silken thread is stretched tight. Holding it in this position, he jostles and nudges the female so that she eventually stands directly underneath with her body parallel to his line. And then she reaches up with her abdomen and picks off the droplets, one by one.

◇

Hunting animals have a particular problem when it comes to mating. On land, a pair has to get to close quarters to exchange sex cells but that inevitably brings them within range of one another's weapons. Somehow or other they have to ensure that mating is not also suicide.

Scorpions are probably immune to their own venom, but individuals could easily wound or even kill one another with their pincers. So it is imperative for an individual seeking to mate, to make it plain that he is a potential partner and not prey. The male initiates events by standing with his eight legs widely and firmly straddled, and juddering his whole body, rocking back and forth. His pectines, those strange sense organs on his underside, are now wide-spread. It may be that he is assessing the character of the ground beneath him for its suitability will become of great importance. The female approaches and he strikes at her with his tail using it as a club with the sting tucked down out of the way. She stands her ground and gives as good as she gets. Now he brings his great pincers forward. So does she. They grip one another. And they begin to dance.

It is a long performance. At its most brief it may go on for five minutes but usually it lasts half an hour or so. It has, in experimental circumstances, lasted for two days, though that may have been because the male failed to find the kind of ground beneath his feet that he required. As it proceeds, he brings his tail forward and this

time he stabs the needle of his sting into the soft membrane in the joint between the two claws of her pincers. This does not kill her. It seems instead to tranquillise her and make her more acquiescent. As they move back and forth, so their faces, if indeed a scorpion can be said to have a face, get closer until they meet and the male kneads her mouthparts with his.

By now, he will have found the surface that suits his purpose. He extrudes a tiny spike which sticks upright on the ground. This is a capsule containing sperm. The precise shape varies with the species. He then starts to jerk backwards, dragging the female with him until she is so placed that her genital opening is directly above the spike. Sometimes, to make the transfer easier, she will tip forward, as though standing on her head. Under the press of her body, the spike bends and as it does so, two tiny valves open, releasing a bundle of sperm which the female takes up into her genital opening. The two then break away.

The scorpions have a group of relatives that so far have not been mentioned. They are no bigger than an average house

△
Scorpions, when mating, clasp one another's pincers, ensuring that neither can damage the other with them.

spider, but they are certainly not spiders, for although they have eight legs, like spiders, their first limbs are armed with pincers and they do not have a spider's thread-like waist. Instead their body is broad, like that of a shortened miniature scorpion. They are harvestmen. Most of the five thousand or so species of them live in the tropics, in South America and Southeast Asia, but a particularly long-legged form occurs in Europe and commonly appears in considerable numbers in the autumn. They, like their scorpion cousins, are aggressive hunters and eat slugs, bristletails and other small creatures. Some species are even strong enough to break open a snail shell with their pincers and eat the occupant.

They and they alone among these early segmented pioneers, have developed a device that enables a male to transfer his sperm directly into the female's body. He has a penis. It is of inordinate length. Indeed, when everted, it can be three and a half millimetres (about an eighth of an inch) long. That may not sound much but it is in fact longer than his body. In most species there is virtually no courtship. The male simply runs at a female, straddles her and copulates. The whole process is completed in about two minutes.

A Central American species of harvestman. Their eight legs reveal that they are related to spiders not insects.
∨

But in one exceptional and extraordinary species, mating matters have become very much more complicated. This harvestman, *Zygopachylus*, lives on tree trunks in the tropical forests of Panama. The male collects mud and bits of bark from the tree trunk, moulds them into little pellets and with them builds a circular arena about three centimetres (about an inch) across surrounded by a low wall. It takes him about two days. He cleans the inside floor, carefully removing any strands of fungus that might grow across it and fungi are very abundant everywhere in this humid warm environment. If rainfall damages the walls, then he carefully repairs them.

It is here that females come to visit him. One will clamber over the wall and start to tap the floor of the arena with her first pair of legs. Male and female face one another and she taps him – though this time with her second pair of legs. The process is a long one. A female may stand near by the proprietor of the arena for an hour or more, motionless except for the very occasional taps she makes on his carapace. Eventually the male reacts. He may abruptly lunge at her and bite her in a sensitive place such as the soft membrane in the joints of her legs until she retreats from the arena. But more usually, he will accept her and signals that he has done so by tapping her with his front legs. He begins to move around her, tapping her all the while. Finally the two face one another. The female grabs the front of the male's body and pulls him towards her. The male everts his penis and inserts it into her genital pore and inseminates her.

Within twenty minutes, the female starts to lay her eggs. The male moves beside or behind her and starts alternately tapping the female's body and the floor of the arena. She raises herself on her long legs and the male may creep beneath her, still tapping the arena floor. She then bends her legs so lowering her body, and extends a long thin tube from her underside and deposits a fertilised egg on the floor. She carefully treads around it with her first legs, kneading the surface, so that it becomes half buried. This performance may be repeated once or twice and then the female leaves the arena.

But the male remains, ready to welcome another female should she arrive. And other females may well do so until eventually the male has up to a hundred eggs of different ages in his arena. He guards them with great care. Wandering males may arrive. If they get a chance, they will eat the eggs, but he wrestles with them, claw to claw, and usually manages to chase them away. Ants too are a danger. Occasionally a lone individual, scouting around for food, will wander into the arena. The male will pick it up with his pincers and, literally, throw it out – though if a whole swarm of ants arrive he may have to retreat and abandon his charges. Few of the first colonisers of the undergrowth can excel him in the care with which he selects his mates and the trouble he takes to guard his offspring.

◇

By three hundred million years ago, large parts of the Earth had become green. The flowering plants had not yet evolved and would not appear for another 130 million years. But some of the giant club mosses, tree ferns and horsetails rose to heights of 40 metres. There were still no large backboned animals roaming through these trees. Some large four-legged creatures were, however, beginning to haul themselves out of the water. They were the first amphibians and their descendants – reptiles, birds and mammals – would eventually dominate the land. But for now, the smaller creatures without backbones had it largely to themselves.

Two groups among them would give rise to populations which would eventually eclipse all these early forms in both numbers and variety. One of these groups was related to the scorpions. They were the spiders. The others were the descendants of the little bristletails and springtails. They were the insects, who today constitute about three-quarters of all living animals on earth. So far nearly a million of them have been described. There are probably three or four times as many still unrecognised and without names. They are, today, by far the most successful and varied of all invertebrate groups.

2

The first to fly

Insects were the first animals to colonise the air. The evidence is clear and incontrovertible. It is a dragonfly's wing, preserved with marvellous perfection in a block of limestone that was laid down 320 million years ago. The first flying backboned animal of which we have any knowledge was that most famous of fossils, *Archaeopteryx*. It had the body of a reptile with a bony tail and jaws laden with teeth but it also had feathers and it dates from 180 million years ago. So insects were in the skies a hundred and fifty million years before birds.

Although the abdomen of the fossil dragonfly is missing, the single wing is virtually complete and careful measurements make it clear that its owner had a wingspan of at least 73 centimetres (nearly 2 ½ feet). This dragonfly must have been the size of a seagull. So the early flying insects, like the early crawling millipedes, developed into giants. Why should this be? Various suggestions have been made. We know that the level of oxygen in the atmosphere at this period was much higher than at present. Maybe this supplied the extra energy needed to power such big bodies. Perhaps it was simply that there were few predators to keep such pioneering animals down to size – no birds in the skies to swoop on a lumbering dragonfly, no lizards to snap up an ungainly millipede. We can only guess.

But it is easy to imagine the advantages that flight could bring. By this time, new dangers had arrived to threaten land-bound invertebrates. The first amphibians, huge creatures with lizard-shaped bodies and moist skins, were ambling through the forests. The

The wing of an early dragonfly dating from 290 million years ago. Its owner would have had a wingspan of about 30 centimetres (12 inches). Although it is not from the biggest of these early dragonflies, this specimen is one of the most perfectly preserved.

ability to flutter into the air in front of them could have been a life-saver for an invertebrate.

The forests had also changed since the first scorpions and millipedes had ventured into them. There was now food to be found in them, not only on the ground but high up among the branches of the great horsetails and tree ferns. Up there, sap could be sucked from stems, young shoots could be nibbled and predators could find more prey. So flight would certainly have brought benefits. But what were the evolutionary stages that led to the development of wings?

We have little evidence as to what kind of flying insects preceded the giant dragonfly but it is clear from the perfection of the wing itself that it must have been the product of a long process of evolutionary refinement. And we can at least see how an insect today manages to repeat that transition from crawling to flying.

The necessity to moult that came with the acquisition of an external skeleton also brought the opportunity to change shape in a radical way. The first terrestrial invertebrates, as far as we know, did not take advantage of it. A baby millipede when it moults becomes a slightly bigger millipede. A young scorpion is simply a miniature version of the adult. But a young dragonfly for the majority of its life is nothing like its adult parents.

It hatches underwater from an egg that, in many species, had been deposited in a slit in the stem of a water plant. The young creature that emerges has, like all insects at some stage in their

lives, a body that is divided into three sections: a long segmented abdomen, not unlike that of a bristletail; a middle section, the thorax, with three pairs of legs attached to it; and a ferocious-looking head, with a pair of huge multifaceted eyes and large mouthparts that are normally folded in front of the face and are accordingly called a mask. It breathes by means of gills that are enclosed in a chamber at the end of its abdomen. It can expel water from this gill chamber in a sudden jet so that it can dart forward. As it does so, it extends its mouthpart-mask into a pair of forceps with which it grabs its prey – an aquatic worm, a small crustacean, a tadpole or even a small fish.

This strange aquatic monster may live underwater in this form for up to six years. During that time, it may moult as many as fourteen times, on each occasion becoming a larger version of its previous self. Eventually, however, it prepares for a final and very different transformation.

It usually starts the process either in the evening or early morning. Jerkily, it climbs up a reed or a rock. When it is well clear of the surface of the water it stops, head upwards, and takes a firm foothold to steady itself. After a short while its yellowish skin splits, first along its head and then, as the creature within hunches itself, along the back of its thorax and down its abdomen. Laboriously, the insect begins to haul itself out. Its body is soft and flaccid and as it draws out the last of its legs, it flops backwards so that it hangs down, head first, attached only by the lower portion of its abdomen that is still within its larval skin.

After a short wait during which it seems to be gathering its strength, it suddenly flexes its body, rears up and clasps the larval skin with its legs so that it clings parallel to its old armour. Then with a final effort it pulls out the remainder of its abdomen.

There are two brown swellings on its back. They are crumpled folded wings. Liquid is pumped along veins within them so that they slowly distend and lengthen. At first they are cloudy, but as they increase in size, the membrane between their veins becomes more tightly stretched and they become transparent. The abdomen, which until now was short and fat, starts to lengthen and thin.

◁

A dragonfly larva is as ferocious a predator underwater as it is in its adult form in the air. This one has seized a tadpole with its mask-like mouthparts.

◁ *A dragonfly
larva or 'nymph'
leaves the water,
hauls itself up a
twig. Half out of its
larval skin, the
body is still soft and
flops backwards. It
gulps air, inflating
its abdomen and
extending its still-
closed wings. Its
body hardens and its
wings open for the
first and last time.*

*A dragonfly prevents
itself from
over-heating in the
sun by adopting this
'obelisk' position,
with its abdomen
pointing towards the
sun.*
▽

It holds its wings closed above its back. They are not yet coloured, nor will they be for a day or so. The liquid that made them
expand is now withdrawn into its body. Then, abruptly, it opens
its wings and spreads them horizontally. It will never close them
again. The dragonfly is ready for its maiden flight.

◇

The dragonfly's wings can be seen in the larvae, if you look
closely. They are there as small brownish seemingly useless outgrowths on the top of its thorax. But why and how did they
originate ancestrally? What function could such structures have
served when they were still too small to operate as wings? Perhaps
they began as flanges on the thorax that helped an insect in regu-
lating its temperature. Spread to the sun they might have gathered
heat when it was needed. Held together vertically above the back
they would have lost heat when that was necessary. Or maybe initially, since they can develop bright colours, they were used as

signalling devices, helping to attract mates or scare away rivals. Perhaps if they developed first among tree-living insects, the initial advantage they brought was to steady and prolong an individual's passage through the air, as it jumped from branch to branch. All three functions have been suggested.

But there is one more possibility. The presence of muscles and a joint at the bases of the wings proves that proto-wings were not simple outgrowths of the thorax but rather an adaptation of some kind of pre-existing appendage such as a gill. There are such things. Other ancient insects, the stoneflies which, judging from their anatomy and from fossil evidence, are thought to have evolved soon after dragonflies, also have aquatic larvae. Some species have gills sprouting not only from the tip of their abdomen, and from their necks – but from their thorax. When they reach maturity, the larvae crawl to the edge of the river and there change into the adult form. And where once there were gills on the thorax, there are now two pairs of wings that lie neatly folded over one another along the insect's back.

Stoneflies may well represent an early stage in the evolution of insect flight. Even though they have wings they seldom fly far. Most of their adult life is spent crawling among the stones beside a river, grazing on algae or lichens and picking up pollen grains. When they do eventually take to the air, their flight is only feeble. They simply stretch their wings and allow themselves to be lifted from the ground by little gusts of wind. This enables them to skim over the surface of the water to escape from predators beneath and to travel in search of fresh feeding grounds or mates.

Dragonflies are much more powerful aeronauts. Their wing muscles are very big indeed. A dragonfly at full speed can travel at 65 kilometres (40 miles) an hour. The two pairs of wings, which are very large compared to the weight of the body they lift, can be beaten independently. That ability enables dragonflies to be very manoeuvrable in the air. They can hover, fly backwards, shoot vertically upwards or downwards, and turn within their own body length. They are hunters. They can hold their six legs in front of their mouth so that they form a basket and with this they scoop up

A dragonfly, like this southern hawker in England, is able to beat each of its four wings independently.

flies and other aerial prey. Some of the larger species are even capable of gathering up small frogs.

For the first two or three weeks of their adult life, they range widely, sometimes flying several miles from the waters from which they emerged. During this time, they perfect their aeronautical skills. Their external skeleton hardens and their colours intensify. Eventually they will try to establish a territory for themselves over a patch of water.

To do this a male will have to fight. These duels can be very violent. Two rivals dart and dive at one another. Sometimes one will knock the other clean out of the air and down into water. A good dragonfly territory must contain an adequate population of insect prey to feed the male himself but also to attract a female. And it must, in addition, provide a suitable site for a female to lay her eggs. For some species, the possession of water will be quite enough, for when the females lay they merely drop their eggs in the water. Others require water plants down which they can clamber to insert their eggs in the stem beneath the surface. Having established his domain, a male will patrol the air-space above it either by flying almost continuously or by stationing himself on a reed or a nearby post from which he can make regular circuits.

His mating technique rivals that of ground-living invertebrates in its anatomical complexity for, he alone among flying insects, does not transfer his sperm directly into the female. The duct from his sperm-producing glands, his testes, opens on the underside of his ninth abdominal segment, near the tip of his abdomen, just where you might expect it to be. But it has no structure with which to transfer the sperm to a female. So before he mates – and this can happen either before courtship or during it – he loops his abdomen forward and transfers his sperm into a holding device on the underside of the second or third segment of his abdomen, just behind the thorax. This is quite a complex organ, consisting not only of a receptacle to hold the sperm but an instrument with which to deliver the sperm to the female – a rod with a bristling collection of spines around its tip. It is termed scientifically a penis, even though it is not, like the penis of a vertebrate, connected directly to the testes. For reasons that will become clear, it might be more aptly called a grouting tool, a bottle-brush or a ramrod.

The male's reaction, when a female enters his territory, varies with the species. In some, he simply flies directly at her and grabs her in mid-air with his legs. Then, as they fly, he arches his abdomen forward and fastens the claspers that it carries on its tip around her neck. Sometimes he grips her so violently that he may inflict real damage. Occasionally he pounces with such speed that he grabs a female of a different species and has to release her when she protests violently and he discovers his mistake. Once his hold is secure, he lets go with his legs and straightens his body. The two are now flying in tandem, the male leading, the female trailing.

In other species, the male conducts a more elaborate courtship. He flies to a spot in his territory that is suitable for egg-laying – a place perhaps where the leaves of aquatic plants break the surface

A pair of darter dragonflies flying in tandem, the male with his claspers firmly clamped on the female's neck.

of the water – and calls a female's attention to it by hovering over it and slowing his wing beat. She then flies towards him and slows her wing-beat in a similar way. Only then does he grab her.

Once the two are flying in tandem, the female swings her abdomen forward so that the genital opening at the end reaches the male's sperm store. The bodies of the two now form a wheel. If they continue to fly the connection will be comparatively brief. But often they will alight and then they may remain together for over an hour. During this time, the male's abdomen can be seen to be pulsing. But this is not because he is transferring his sperm. Instead, he is eliminating any that might lie in the female's duct as a result of previous couplings.

Different species do this in different ways. Those with bristles on their penis use it to scrape the duct clean. Others have an inflatable bulb at the end and with this they ram any sperm that might be in the oviduct from a previous mating into the upper-most part of it. Eighty percent of the time that the male is connected to the female is spent in this activity. If the female shows any sign of restlessness during the process, the male may

69

◁

A dragonfly pair in the 'cartwheel' position. The male is still clasping the female at the back of her neck, while she has curled her abdomen forward to collect his sperm from the pouch just behind his thorax.

slap her around the head with his rear pair of wings. Only after it is completed does he inject his own sperm.

In some species, the male will maintain his grip on the female's head while she lays her eggs. He continues to protect his genetic inheritance, even after he has parted from the female by keeping her in his territory and flying close by her ready to drive off any other male who might court her. Yet in spite of all these efforts, only fifty per cent of males succeed in fathering offspring. All the females, on the other hand, are fertilised.

Damselflies are very close relatives of the dragonflies. They differ from them in having a much more slender body and, when they perch, in holding their wings closed above their backs instead of keeping them outstretched as a dragonfly does. There is also an easily identified difference in their larval form. Damselfly larvae have external gills whereas dragonfly larvae do not.

Damselflies courting above the surface of a pond.
▽

◁
A male Agryon *damsel flaunts his handsome wings, inviting a female to allow him to copulate.*

They are particularly enterprising in finding water in which to lay their eggs. Cascade damsels lay their eggs on the wet rocks behind the main veil of waterfalls. The larvae, when they hatch, have to hang on to the rocks to prevent themselves being carried away and to grab their prey with lightning speed from the water as it sweeps by them.

Megaloprepus, a damsel that lives in the rain forests of Central America, uses the tiny pools that accumulate in tree holes. All damselflies are feeble flyers compared to their cousins the dragonflies, but *Megaloprepus* has a wingspan of 20 centimetres (nearly 8 inches), the biggest in the entire damsel-dragonfly group, and it does not have the strength to flap its wings at any great speed. The males, having found a suitable pool, guard it by circling above it and beating their handsome blue-blotched wings in a mannered ritualised way that gives them the popular name of 'helicopter' damselfly. The display not only keeps other males away but also attracts females anxious to mate and lay.

The Sabino Dancer damsel breeds in creeks in the Arizona desert. The adults emerge from their pupae just before the late summer rainy season when water levels are at their lowest. When the storms come at last, there are flash floods and the levels of the creeks may rise by as much as a metre or more. As the skies clear, the Sabino Dancers begin their courtship. But they face a serious problem. It is so hot that water quickly evaporates and the level of the streams may fall by as much as 5 centimetres (2 inches) a day.

◁
The males of some damselfly species merely fly above their females after mating to ensure that no other male usurps them. Here, Platycnemis *males are even more protective of their paternity and keep hold of their females until have finished laying.*

If a female were to lay her eggs close to the water surface, they would soon be left high and dry. So the female Sabino Dancer, having copulated and with the male still clinging to her back, climbs down the rocky flanks of the stream bed until she is as much as a metre below the surface. There she lays her eggs, inserting them in the stems of water plants, as most damsels do. She and he may both remain there for up to an hour, breathing the air that has been trapped by hairs along their bodies and beneath their wings. But the hazards of breeding in these desert streams are formidable and today only a dozen or so individuals still survive.

Damselflies, dragonflies and stoneflies are considered to be among the most primitive of flying insects. We cannot assume that their long-extinct ancestors behaved in exactly the same way as their descendants do today, but the fact that they are both, in their larval stage, aquatic suggests that the spur to flight might have been something other than to escape enemies or to seek food in the tree tops. It might also have been a way of extending their search for a mate beyond a short stretch of river-bed. That certainly is the case for another group of early flyers, the mayflies.

Mayflies are found in freshwaters throughout the world. There are about three thousand different species of them, but one of the biggest occurs in a few rivers in central Hungary. Unlike many mayfly species which emerge from their aquatic larval life at different times throughout the warm summer months, these Hungarian mayflies emerge simultaneously in vast swarms on just a few occasions in midsummer.

The day before they do so, there may be little if anything to suggest the scale of the swarming to come. The river, fifty metres or so across, slides quietly between its willow-lined banks. A kingfisher perches on a low branch beside its nest hole. A wagtail hunts for mosquitoes in swooping circuits. Dragonflies hawk up and down among the reeds and male frogs stutter their honking mating calls. A couple of local fishermen drift downstream in their dinghy, occasionally striking the surface of the water with a cone-shaped instrument that makes the dull thudding noise that they believe attracts the carp that they seek. But they have a resigned air, for they know that the mayfly emergence is due any day – and then their fishing will be pointless. When that happens, the fish in their river will have so many insects to feast on that none will be tempted to take their hooks.

As afternoon cools into evening, dimples appear on the river surface. You can see, through binoculars, that they are made by large pallid insects pulling themselves free from their larval skins which are left floating on the surface. They have a pair of long filaments trailing from the back tip of their abdomen. That tells you they are mayflies. But you will have to be quick to spot them for

they are not likely to be there for long. Either there is a splash as a fish leaps from the water to snap them up before they have even left it or the kingfisher will have flashed down in a blue streak to collect them. But within minutes, the surface of the river is pock-marked with dimples as thickly as if there were a rainstorm.

All the individuals now emerging are males. Soon there are so many that neither fish nor fowl can collect them. They flutter towards the river-banks and alight in droves on the trees, covering the branches like some newly-sprouted crop of fruit. They are so eager to find a perch that they will gratefully alight on your hand if you offer it them. For they have one urgent job to do.

Mayflies having acquired their wings once, now, strangely, moult a second time. As a male sits on your hand, with his wings held vertically above his back, his soft limp abdomen begins to pulse. The skin splits and within seconds, the adult mayfly starts to pull himself free, dragging his legs out from their tubular casings,

75

and leaving them behind like thin wrinkled stockings. His new incarnation differs only slightly from his previous one. His tail filaments are marginally longer, his abdomen glistens more brightly and his wings, which in his sub-adult form were dull brown, are now a shiny transparent blue. Why mayflies should have these two adult forms, no one knows.

The males have no time to waste. The females are now rising from the surface of the river. The males fly off almost immediately to find them. When one reaches the surface, a dozen or so males may descend upon her in a jostling scrum. As soon as one succeeds in copulating, she manages to fly away and the males resume their search for another virgin female. By now the blizzard of insects is so thick that it is difficult to see across the river to the opposite bank. The males continue to compete in finding females, swooping and fluttering in vast squadrons over the water. They have bifocal eyes to help them in their frantic search for a mate. The upper part of each is enlarged with bigger facets that allows them to spot any female in the air above them. Neither male nor female feed. Indeed they cannot, for they have no mouth and their tiny

△
Male mayflies cruise up the river in search of unmated females that may have just emerged on the surface of the water.

stomachs are filled with nothing but air. Within the hour, successful or not, they are all dead, and their bodies accumulate in thick
curds on the river's surface.

The females, however, live a little longer for they have one further task. Now that they have been fertilised, they must lay the
several hundred eggs that each carries in her abdomen, almost
completely filling it. Their method is simply to drop them on the
surface of the water and allow them to drift down to the bottom.
But these eggs take forty-five days to hatch. If they were laid on
the stretch of the river where the females emerged, then the larvae
would hatch much farther downstream, having been carried by
the current. In time the whole population would end up in the
sea. To compensate for this, the females now fly upstream. They

*The annual rise of
the Hungarian giant
mayfly has ended
and a frog feasts on
the bodies of the
dead males floating
in the shallows.*
▽

travel in long processions, seeking a place where the water is deep
enough and the current has the right speed, to carry their eggs
down to approximately the place where they themselves emerged.
There the larvae will find safe homes in the mud and gravel of the
bottom and in due course they will emerge from the same stretch
of river as their parents did.

Not all mayfly species behave in this way. Some emerge in smaller numbers but at more frequent intervals throughout the summer. Some spend only a year as a larva, some as long as four. And some adults, instead of living only a few hours may survive for several days. In a few species, the males assemble in aerial columns several metres high, into which the females fly to be mated. As a male seizes one with his forelegs, the two tumble down through the air before the female separates herself and flies off to drop her eggs into the river. But no mayflies feed during their short adult lives. Flight for them is simply the road to copulation.

Grasshoppers use their powers of flight for very different purposes. The group is not quite as ancient as the dragonflies and mayflies but nonetheless it dates back to around 260 million years. The young larvae that emerge from the egg live on land, not in water. But like larval dragonflies they do not, in their first incarnation, have wings. When, after several moults, they emerge as adults, they like dragonflies are equipped with two pairs of wings. These operate in very different ways. The front pair is relatively narrow and leathery. The back pair, which are normally kept folded like a fan, are wider, thinner and often coloured bright red or blue. Grasshoppers feed on leaves, chewing them vigorously with their scissor-shaped mandibles. Normally they move from stem to stem by jumping with their long and powerful hind legs. Only when they are alarmed do they take to flight. Then they fly away explosively, in an arcing flash of colour from their hind wings. Your eyes follow them as they rocket through the air but then – in mid-air – the colour suddenly vanishes. You follow the line of their leap with your eyes and may think you can judge where they must have landed. But usually you are mistaken. They dropped to the ground immediately they shut their wings and landed much nearer to you than you expected.

However some members of the group, the locusts, use their wings not for short-range escapes but for extremely long-range journeys. There are at least ten different species of locust but they are all around 6 centimetres (2½ inches) long. They look like rather large grasshoppers and indeed for much of the time they

▷
An adult blue-winged grasshopper in Sardinia leaps through the air, its tough protective forewings raised, its rear displaying the flash of colour that may momentarily disconcert a pursuer.

◁

A plague of grasshoppers in Panama devastating the vegetation. Being immature they are flightless but they have acrid body fluids and warn off predators with conspicuous colouration.

may behave very much in the same way, living largely solitary lives, feeding on what vegetation they can find on the hot dry savannahs of Africa and laying eggs.

When the young first hatch, they have no wings so they can only travel by hopping and early in the season they tend to hop away from one another with such vigour that they become widely dispersed over the land. Little vegetation in the area escapes them. But as the adults continue laying and their numbers of young grow, a change takes place. If the juveniles are so crowded that they happen to touch the upper part of one another's legs during the few hours before they lay, then the females will add a special substance to the foam with which they surround their eggs.

The young locusts that now hatch as a consequence of this inoculation are very different. They are not green like their parents and older siblings but dramatically coloured with stripes of black, yellow and orange. Indeed, they are so unlike their predecessors that until not long ago, they were considered to be a different species. And they also behave very differently. Instead of avoiding one another, they gather together in bands and travel and feed together, moulting and growing and consuming everything in their path that is green and edible.

At their fifth moult they acquire wings. The whole swarm, having eaten all the vegetation within hopping distance now takes to the air. A few individuals begin the departure, taking off, spiralling upwards and than descending again to the ground. Increasing numbers do so until eventually the whole swarm takes off. The numbers of individuals it contains may be vast – as many as fifty billion individuals. Together they form a dense cloud that may cover hundreds of square miles. Carried by the wind they can travel great distances. Even without that help they are able to move at speeds of up to 15 kilometres (10 miles) an hour. Swarms of these dimensions can cause national disasters, consuming hundreds of tons of vegetation a day, destroying crops and bringing famine to humanity as they have done since biblical times.

◇

▷ ▷

A choking, blinding swarm of locusts sweeps across Senegal, consuming every edible plant in its path.

Many species of grasshopper have an additional use for their wings. They use them as musical instruments. Each of the robust front pair has a prominent vein which they rub with a row of stout chitinous pegs along the inner surface of their long hind legs. This creates a loud buzz, the very voice of summer. In most species, only the male sings. It is his way of asserting his territorial rights and attracting a female. Her ears are thin circular patches on either side of her first abdominal segment. Each species of grasshopper (and there are around twenty thousand of them) has its own characteristic song, so a female can tell not only where a male is singing, but also whether or not he is a suitable mate for her.

The crickets, which are close relatives of the grasshoppers, sing even more vociferously and penetratingly and produce their sound using their wings alone. Each of their forewings, on its underside, carries a row of fine teeth along its length like a blunt saw. This is scraped rapidly against the strengthened edge of the opposite wing. Each wing also carries a circular patch which is thin and taut like the membrane of a drum and which vibrates with the sound, so amplifying it. The call can be made even louder if the singer depresses his abdomen to create a resonating chamber beneath the vibrating wing.

The female cricket's ears are on her forelegs – a slit on each of her upper thighs that leads to a little chamber the walls of which are so thin that they vibrate when the sound waves strike them. Each slit itself is so narrow that the little sound receiver has all the directional qualities of a gun-microphone, and as she moves her leg, she can locate the exact direction from which the song is coming.

In North America, the male of one species, the sagebush cricket, gets a strange reward for his efforts. When a female eager to mate arrives, she climbs on to his back. As mating begins she starts to gnaw away at his hind wings. They are particularly thick and provide a reasonable nuptial meal. He even lifts his wings to make it easier for her to get to them and while she is thus engaged, the male transfers his sperm. After that, having lost his nuptial gift, he sings no more.

The mole-cricket's enormous forelegs adapted for burrowing do resemble a mole's fore-paws

Mole crickets have developed a very specialised and ingenious way of amplifying their calls. They spend most of their lives underground feeding on roots, worms and the larvae of other insects. Their forelegs, with which they dig, are sturdy and extremely strong. Their wings, not surprisingly for creatures that spend most of their lives underground, are very short and extend only about half way along the abdomen. Their primary function, however, is not flight.

The European mole-cricket's main burrow has twin entrances which unite a few inches below the surface. A short distance beyond the junction, the tunnel expands into a spherical chamber. It then narrows and continues still further downwards to the compartment where the eggs will be laid. The globular chamber is the male's singing stage. In late summer he sits there, with his head down and his abdomen lifted, scraping his wings together. The two passages that lead from his stage to the surface have been built with extraordinary precision. They widen exponentially as they near the surface. They are loudspeakers which amplify the male's song to such a degree that, on a windless night, he can be heard from nearly half a mile away.

Some sixty million years after the arrival of dragonflies in the fossil record, beetles appear. Today they are by far the most varied group of all insects. So far, 350,000 species of them have been named and no entomologist supposes that there are not as many more still waiting to be recognised. Beetles constitute 40 percent of all known insect species. Indeed, they represent 30 percent of the species of all known animals of any kind. Their variety beggars the imagination. Some are armed with huge horns. Others have antler-like mouthparts. Heads are placed on the end of stilt-like necks that rise four times higher than the body like the periscope of a submarine; snouts are elongated into elephant trunks; antennae, beaded or hair-thin, decorated with tufts or equipped with leaves like a book, may extend for five times the length of the body or be flattened and shortened into squat spoons. Some

The hard chitinous skeletons of beetles can take many forms. The heavy jousting weapon of the male Hercules beetles from central America (below), the iridescent armour of the longhorned beetle from Arizona (opposite above) and the giraffe-like neck of the leaf-rolling weevil from Madagascar (opposite below).

beetles are a metallic gold, others glint emerald green or carry spots of turquoise blue. The smallest beetle is even smaller than the head of a pin. The biggest – such as the titan beetle that lives in the forests of the upper Amazon – can measure 17½ centimetres (6¾ inches) from the tip of its ferocious mandibles to the end of its armoured abdomen. Why should beetles, above all insects, have been so spectacularly successful and exist in such varied forms?

One source of their success may be connected with a simple but fundamental change in the use of their wings. Like the insects that preceded them, they have two pairs. The back two are membranous and used in flight. The front pair, however, have been thickened, hardened and converted into covers to protect the hind wings. So characteristic is this development that it has given the whole family its scientific name – Coleoptera, 'shield wings'.

△
Many weevils, of which there are almost fifty thousand species, have a drab camouflage, but others like this species from Papua New Guinea are brilliantly coloured.

▷
The goliath beetle, from equatorial Africa, is one of the contenders for the title of heaviest of all insects. Specimens may weigh up to 100 grams (3.5 oz).

The fore-wings themselves also have a special name. They are called elytra. In flight beetles usually hold up their elytra out of the way of their beating rear wings. The posture is hardly aerodynamically efficient, so beetles tend to have a rather lumbering, laboured flight compared with other insect aeronauts. Only one group, the rose chafers, which includes the goliath beetle, do otherwise. When they fly, they keep their elytra shut and their membranous wings emerge from underneath the outer edges.

When beetles alight, they face another problem. Most elytra are shorter than the fully extended hind wings. If these hind wings are to be fully protected, then they must be folded before they are packed away. Many beetles perform this feat with an elegance and ingenuity that would do credit to a Japanese master of origami. In flight, the wings are held taut between two main rod-like veins. These veins project beyond the wing joint into the body cavity where they are attached to muscles. When these muscles contract, the veins are pulled apart so that the membrane between them is stretched taut. When they relax, the membrane naturally falls into

90

longitudinal folds. The muscles can then swivel the two wings through ninety degrees so that they rest parallel to the abdomen and can be rolled up.

Some beetles fold them in an even more complex manner. The secondary veins of their wings have spring-locked mechanisms which work in much the same way as they do in a pop-up book. When these unfasten, the wings can be folded transversely and so form a small neat parcel. As the elytra close, patches of backward pointing spines on their undersides and on the upper surface of the abdomen engage with the parcelled wings to ensure that they stay tightly folded.

With their delicate flying apparatus fully protected in such ways, beetles can tackle all kinds of rough tasks. They can dig into the ground and barge their way through rubble. They can squeeze into tight cracks, bore into wood, tunnel into dung and even, by holding a bubble of air beneath the elytra, swim.

Ancestral beetles made this invaluable and pioneering change in the function of their forewings at a very early stage in insect history, a mere sixty million years after the dragonflies took to the air. That gave them a head start over other kinds of insects that were to evolve later. They were thus able to diversify into all the varied forms needed to exploit the opportunities that awaited them on the ground. Few other kinds of insects have since managed to displace them.

The diversification of beetles was aided by another development. Dragonflies change from larval to adult forms in the course of a single moult. Beetles, however, take a long pause between the two states during which they rest, sometimes for months, as a pupa. The name is very appropriate for the pupae of many beetles do indeed look like little dolls, which is what the Latin word means. They lie, often in special chambers or in cocoons, encased in a sheath of chitin through which sometimes limbs and eyes, antennae and wing buds can all be clearly seen, beautifully coiled and packed. But some beetles, ladybirds among them, have larvae which look like particularly well-wrapped mummies and in them hardly any details can be distinguished.

This intermission in their lives enables beetles to rest through an unsuitable season. It also gives them the opportunity to reorganise their bodies in a profoundly radical way and the creature that eventually breaks free from its pupal wrappings bears very little resemblance to the comparatively featureless grub. Thus beetles have been able, as adults, to acquire the sturdy compact business-like shape of a scarab, the spectacular antennae of a longicorn five times the length of its body, and the massive medieval jousting equipment of a hercules beetle.

The length of time a beetle spends as a larva varies greatly. Ladybird larvae, ferocious little creatures which prey on aphids and other insects, usually moult three times, pupate and emerge as an adult all within a period of about four weeks. By contrast, stag beetles, which feed on rotten wood, take a much longer time to get all the nutrition they need from such an unpromising diet. Some very large species spend up to seven years steadily masticating and digesting their food until at last they pupate. But even that is not the longest of larval lives.

Cicadas are the champions. They belong to a group of insects, the bugs, which have mouthparts modified into a thin piercing instrument through which they suck the sap from plants. They are about 4 centimetres (1½ inches) long with flattened bodies, transparent membranous wings that are normally held flat across their abdomen, and piercing voices. In the tropics in the evening, their songs can be deafening. Many sing with such regularity at such precise times that they acquire nick-names such as the 'six o'clock beetle' and are taken by human listeners as a cue for an evening drink. But one species that lives in the eastern United States takes this clock-watching habit to an extraordinary length.

The periodic cicada is a handsome insect with a black body striped with orange, transparent glistening wings and large scarlet eyes. A female lays her eggs in summer, depositing them in slits which she cuts in the twigs of trees. After six to eight weeks the eggs hatch and tiny young cicadas fall to the ground. They are wingless but do have short powerful fore-legs and with these they dig themselves into the ground. There they continue tunnelling

▷
After seventeen years spent below ground drinking sap, a North American periodic cicada sits singing on a branch with empty larval cases still clinging to the leaves beneath.

until they find a tree root. When they do, they latch on to it with their mouthparts and start drinking the tree's sap. They may move from one root to another, should the first root develop some sort of defence, but they remain underground. For seventeen years.

When that time has expired, the larva in its tunnel starts to dig a vertical shaft up towards the surface. Every now and then it returns to its tree root for a further drink and excretes liquid from its rear end which it mixes with the fine dust produced during its excavations. This forms a mortar which it uses to give a smooth lining to its shaft. Sometimes, it will continue building above the surface of the ground and so produce a small turret. And then it waits.

The date of its next move is largely predictable to an extraordinary degree. It may be influenced by temperature and wind by a day or so. But on one particular day in June, the larva breaks the thin roof in the top of its turret, pulls itself out and clambers up the tree from which its egg fell. There it breaks its larval skin and emerges as a full-grown winged insect. What is more, a million other cicadas will be doing so at the same time. Over a period of three or four days almost all the population of seventeen-year cicadas will emerge. How this synchronicity is achieved we still do not understand. The weather may be one cue but that could only shift the date by a few days one way or the other. How do the larvae know that seventeen years have passed? It is certainly possible that each spring they are able to taste a difference in the sap since it will contain a sudden increase in the amount of sugars. But that only puts the mystery one stage back. How do they count the number of springs? We do not know.

This simultaneous emergence brings the cicadas a great benefit. It completely overwhelms the birds and other predators that prey on such large insects. The local birds, however intensively they might feed, can only possibly eat a tiny proportion of the hordes that appear. Within a day or so, they can eat no more at all. And still the cicadas come. They completely clothe the tree trunks. And they sing.

Though the deafening level of the sound might not suggest so,

▷
Newly emerged seventeen year cicadas cover the trunks and branches of the trees on whose roots they have been feeding in the eastern United States.

94

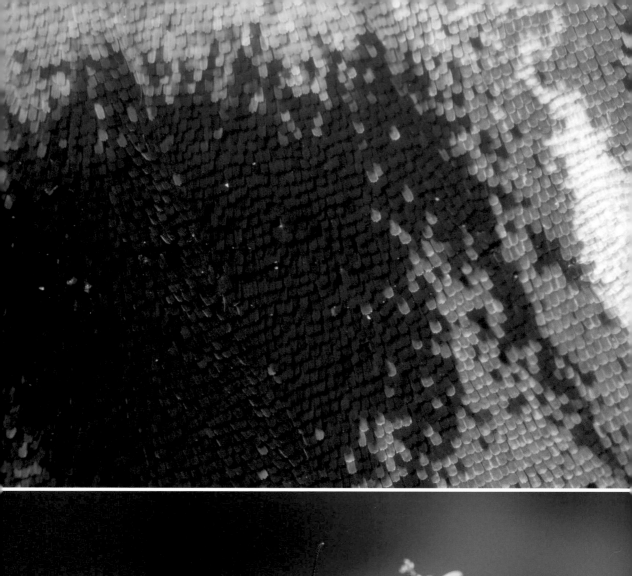

◁

The colours on a butterfly's wing come from rows of tiny scales. Some owe their colour to a chemical pigment. Others create it with an internal structure that refracts light and gives the wing an iridescence.

it is only the males who are singing. The females are all silent. The males produce their song by vibrating a pair of circular ridged membranes, one on each side of their first abdominal segment. This produces a high ringing tone and is amplified by air-sacs which largely fill the abdomen itself, causing it to resonate.

The females are not ready to mate immediately but as soon as they are, they signal their availability with a quick flick of their folded wings. This makes a relatively quiet clicking noise. As soon as a male hears it, his sound changes and he crawls along the branch towards the female. You can demonstrate that his cue is entirely sonic by imitating it with a snap of your fingers. That noise is sufficient to deceive a male and one will land on your arm and crawl along it towards your hand, as long as you continue to respond to the chirping that he now makes with more clicks. When male meets female, the two quickly copulate. The female then goes away to lay while the male resumes his calls in the hope that he may be lucky once again. So the chorus continues unabated for the three or four days that it takes for the entire population to emerge. They may feed occasionally on tree sap but within two or three weeks, all the adults will be dead and the song of the cicada will not be heard in the neighbourhood for another seventeen years.

A glasswing butterfly from the cloud forests of Costa Rica has wings that lack most of their scales and are therefore transparent.

◁

◇

Butterflies and moths, as adults, feed almost entirely on nectar which they collect from flowers with the long tubular mouthparts that they keep coiled like watch-springs beneath their heads. To collect the nectar they need, however, they must be able to fly very efficiently and to spend much of their lives on the wing. So instead of dispensing with the help of forewings, as the predominantly terrestrial beetles have done, the butterflies use the two pairs as a single pair of large aerofoils, sometimes hitching them together with stout bristles.

Such big wings can then be used in other ways. Patterned and coloured they can serve as banners of identity, enabling an individual butterfly to recognise potential mates and rivals. These

▷ ▷

Butterflies gather in swarms at the edge of a Peruvian river, probing in the wet mud with their probosces and drinking its mineral-laden water.

patterns are created by lines of microscopic scales which cover a butterfly's wing in overlapping rows like tiles on a roof. Each is the flattened outgrowth of a single cell and is no bigger than a tenth of a millimetre across. And each carries one particular colour. In some, this is created by a chemical pigment. In others, the colour comes from a microscopic structure of tiny surfaces which refract light and in some cases produce an iridescence. Wing scales are the unique possession of butterflies and moths and give the whole group its name – Lepidoptera, scale wings.

The minimum number of patterns needed for a butterfly to identify itself is clearly related to the number of species likely to occupy any particular area – a few dozen. The fact is, however, that there are around a hundred thousand species of butterfly worldwide and each has its own distinctive wing pattern – which an entomologist can, in most cases, use to distinguish it from any other species, no matter what part of the world it comes from.

Brilliant displays, of course, can be dangerous for they will attract the attention of predators such as birds which, by the time the Lepidoptera appeared, had established themselves in the skies in some numbers. Accordingly butterflies that are active during the day, tend to perch with their wings closed above their backs, so that their brilliance is concealed. On the other hand, moths are for the most part (though not exclusively) nocturnal. Bright colours can serve no purpose in the dark, so the wing scales of most moths are used to create camouflaging colours and patterns. Many moths therefore keep them open across their backs as they rest through the daylight hours, invisible against bark or on the underside of leaves.

But butterflies and moths, which seem to have such frail bodies and fly relatively feebly, use their wings to make much longer journeys than one might guess. The cabbage white butterfly, for example, which is such a common sight in a European vegetable garden, seems to be doing little other than indulging in fluttering courtship flights with the females occasionally dipping down to settle briefly on a cabbage leaf to lay a batch of eggs. But the likelihood is that these are not insects that have hatched locally. They

▷
Not all moths are drably coloured. This Madagascan moon moth, the largest of its family, though primarily nocturnal, is nonetheless beautifully coloured and ornamented with long tails to its wings.

may have already travelled many miles. If it is early summer, then most will have come from the south-east. After apparently dawdling around the cabbage patch they disappear. Most will be continuing their journey in a generally north-westerly direction looking for more places to linger and lay.

An adult cabbage white, like most temperate region butterflies, seldom lives for more than three or four weeks, but their eggs laid early in the season can produce another generation of adults long before the summer is over. If the next generation appears before mid-summer, they will continue wandering in a north-westerly direction. But then, early in August, the behaviour of the cabbage white changes. The trigger, it seems, is the increasing coldness and length of the nights. Now the adults reverse the direction of their flight and start to travel back towards the south east.

Monarch butterflies in North America, have taken this travelling habit to extremes. As caterpillars, they feed on milkweed, a plant with such a poisonous sap that few insects can tolerate it. The monarchs, however, have evolved an immunity to the poison and accumulate it in their bodies. They themselves therefore are poisonous and most birds leave them alone – and that has an important bearing on their subsequent behaviour.

In any one season, as many as five generations may live, breed and die in North America. But as autumn approaches, many of those reaching maturity start to behave in a different way. Although they are adult they do not mature sexually. Instead of developing their sex organs, they start to store fat in their bodies. By mid-September, it constitutes about a third of their body weight. And then they start to fly southwards. As they go, others join them. Soon steady processions are moving over broad fronts along well-established routes. A hundred million of them are now on the move. Each night they roost in trees that their predecessors have used for decades. They feed on nectar from autumn flowering plants as they travel. Sometimes they fly high, taking advantage of strong southerly winds that carry them several miles a day. On they go down through Texas until at last they reach Mexico. There in less than a dozen small sites in secluded valleys

▷

Millions of monarch butterflies take refuge from the North American winter in a few remote valleys in Mexico, covering the branches of the trees.

they roost in fir trees. A single tree may hold many hundreds of thousands. They are so thick that the leaves on some branches are totally hidden by the tightly packed bodies. Their combined weight is so great that some of the smaller branches will break and those that had settled down on it fly agitatedly around the tree trying to wedge themselves into other well-settled ranks.

Were it not for the milkweed poison in their bodies, they would represent an irresistible feast for all kinds of local birds. As it is, they hang there unmolested. Occasionally a shaft of sunshine may play over a laden branch and a few will detach themselves and flutter down to a nearby stream to drink. But otherwise, they remain motionless, conserving their energy.

The fat they accumulated as caterpillars to fuel their southward journey is not yet exhausted. In the spring, they become more active. The great trees on which they had hung throughout the winter now seem to smoke as millions of them take to the air once more. And then over a period of a few days, they all disappear from the Mexican mountains and start on their long flight north. They cannot stay in Mexico because no milkweed grows there. By this time, many will be getting on for nine months old. They fly in a much less determined way, and when they reach the southern shores of Texas it is spring and at last some find milkweed plants on which to lay their eggs. Their journey is over and they die. Others will continue farther north to lay. Soon there are several generations leap-frogging one another throughout the spring and summer until eventually some will reach northern Canada where their forebears, several generations earlier, had started the previous September. Some individuals may have flown 4000 kilometres (2,500 miles).

Migrating butterflies may fly farther than any other insects, but they are not by any means the most skilful insect aeronauts. To fly in all directions, backwards, forwards and sideways, to move at great speed and hover motionless precisely positioned in three dimensions, requires more sophisticated instrumentation. Flies

*Crane flies, or
daddy-long-legs as
they are known,
have particularly
conspicuous halteres,
the modified hind
wings which provide
their owners with
flight information.*

acquired that by ceasing to use their hind wings to generate aerial motive power and converting them instead into sense organs. They have been reduced to a pair of tiny rods called halteres which have knobs on the end like drumsticks. These swing up and down at the same speed as the forewings but out of phase with them. It used to be thought that this gave the fly stability in the air, as a flywheel or gyroscope might do. That is not however, the case. The halteres have microscopic sensors on their upper and lower surface near the base which detect variations in the stresses playing on the beating rods as they react to the fly's path through the air and to air currents that may otherwise cause the fly to pitch, roll or yaw.

The forewings themselves are not driven by muscles attached to internal projections as they are in dragonflies. Instead the flying muscles are fixed to the rigid walls of the box-like thorax. This is constructed from highly elastic chitin which snaps in and out as it is pulled by the muscles. The wings are hinged between the top and side plate of the thorax in such a way that, as they approach the mid-point of their beat they suddenly flick to complete it. This ingeniously simple mechanism enables wings to be beaten as many as a thousand times a second.

A blowfly, by no means the most accomplished aeronaut in the family, is able to land with great precision upside down on a ceiling. It flies up towards it at a shallow angle with its forelegs held stiffly in front of it. As they strike the ceiling, they absorb some of the shock, slowing the fly. But there is still enough forward force acting on the fly's body to carry it forward. It swings on the claws of its front legs and plants the other four on the ceiling so that it finishes upside down and facing in the direction from which it came – all of which requires perfect control and timing.

Hover-flies are even more skilful. One can hang in the air in front of a flower with its wings moving so fast they are invisible to the human eye. Its body is tilted slightly backwards so that the air-flow from its wings is directed vertically downwards with a force that exactly matches its weight. As a consequence the fly itself seems to be perfectly stationary. Then suddenly it vanishes. It has moved away so swiftly that your eye has been unable to follow it and taken up a stance somewhere else.

A hover-fly hanging motionless in mid-air in front of a flower. Although their larvae include some that are carnivorous and attack aphids, the adults feed on pollen and nectar like bees and are important pollinators.
▽

Hover-flies
copulating, England.

Skilful flight is a crucial element in hover-fly courtship. A male takes up an aerial territory located, it might seem to a human observer, in a quite arbitrary way – beside a bush, a gap in a hedge, a bend in a path. He waits for a female to visit him. If she does, he will pounce on her and copulate in mid-air. So eager is he to do so he will chase anything that invades his air space moving at the sort of speed of a female. Even an accurately aimed pea from a pea-shooter will send him off on a fruitless dash to intercept it, just in case it was a real opportunity.

The males of many species of dance flies join together to fly in columns above similar swarm-markers. They are predatory. A male attracts a female with a gift, usually the body of another insect. It may be a moth, a mayfly or another kind of fly such as a crane fly, though these are rather larger than the dance fly itself and take some subduing. Occasionally, a male will even grab a

rival to serve as a nuptial gift. After alighting and probing it repeat-edly with his proboscis, a male flies off with it to his display ground, carrying it beneath his body with his middle pair of legs.

Once there he begins to fly in a very mannered way – in cir-cles, or upward spirals, or from side to side. This is his dance. Females are quick to notice it. They are flying some distance above at the top of the swarm and quickly descend to the male to take his present from him in midair. But he does not let go of it completely. Instead, the two holding the gift alight on a nearby leaf. The female now begins to explore what she has been given with her proboscis. While she is doing so the male transfers his sperm to her. The process may last as long as an hour. But even in that time, she may not have consumed her gift entirely. The male may keep one leg on it and when he has finished, he retrieves the gift and returns to the swarming place to offer it to someone else.

In some dance fly species, the male wraps his offering in tissues of

△
A male dance fly copulates while his female investigates a nuptial gift of silk that probably contains nothing worth eating.

silk thread that he produces from spinning glands on his front legs. The female when she unwraps it, sometimes discovers that instead of something edible she has been presented with an uneatable seed or even a fragment of down. This does not seem to disconcert her, however, for she will still allow the male to complete insemination. If the gift were once a bribe from the male to persuade her to accept him, it has now become a ritualised formality. One might suppose that it is the thought that counts.

In their larval lives, as maggots, flies are the most nondescript of organisms. They have been reduced to featureless eating machines. Most are legless, eyeless, even headless, nothing but bags of soft moist skin with a pair of small hooks at one end. They wriggle their way through rotting fruit, dung, fungus or putrefying flesh. They macerate what surrounds them, saturate it with their saliva and then absorb it. Eventually, when sated, they pupate and turn into small barrel-shaped capsules. But within those walls, miracles of re-organisation take place and from them emerge superlative flying machines, with acutely perceptive eyes, highly sensitive antennae, glinting transparent wings and elegant delicate limbs. Some have mouthparts modified to serve as spongy absorbent mops, others have converted them into hypodermic syringes of such delicacy that we may be unaware of them as they are inserted in us and take our blood.

The twin constraints of an external skeleton that has to be moulted with growth, and a breathing system that relies on largely unassisted aerial diffusion down microscopic tubes, have prevented flies from growing to great size. Instead they have specialised in miniaturisation and done so to an astounding, scarcely believable degree. Some flies, complete in every anatomical detail of their complex anatomy, with eyes and a brain, sex organs and mouthparts, legs and wings, are scarcely bigger than specks of dust.

Insects were not only the first creatures into the air, but to this day no other organisms exploit the freedoms and opportunities of that world with greater virtuosity than they do.

3

The silk spinners

Silk is an astonishing material. Filaments of it are stronger than steel threads of equal thickness. Some of them can be stretched to nearly three times their length without breaking. Others are so elastic that they will absorb the energy of a fast-moving object, slow it to a standstill and then quickly recover their original length. Chemically, silk consists of long strands of amino-acid molecules, similar to those constituting keratin, the substance of hair, horn and feathers. It is produced by glands which may develop in many parts of an animal's body – its abdomen, its mouth, its legs. It oozes into the gland's central duct as a highly viscous liquid but when it is pulled, by a leg, an arm or even the wind, the tension causes it to solidify into filaments.

Terrestrial invertebrates evolved the ability to produce silk several times in their history and quite independently of one another. The first to do so were probably those multi-legged velvet worms, for the sticky mucus that today they squirt at their prey from their mouths is thought by some to be chemically akin to silk. Investigations are not yet sufficiently advanced to be certain. Those other land-living pioneers, the millipedes, however, certainly produced silk – though they do not seem to have exploited it to any great degree. The only use to which a male millipede today puts his silk-producing powers is to spin a small pad on which to place his sperm when he transfers it from the testes on his second segment to the impregnating device on his seventh. Bristletails use silk too – to make the line on which the males hang their little packets of sperm.

Among insects more advanced than springtails, it is

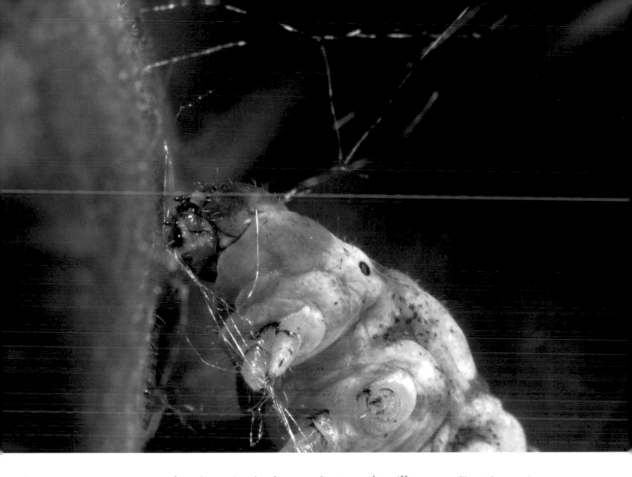

predominantly the larvae that are the silk-users. For them, it serves as a protective wrapping during the vulnerable time when they are changing into their adult form. One insect, an Asiatic moth whose caterpillars feed on mulberry leaves, uses it in a specially lavish manner. The silk-producing glands lie within the caterpillar's head and open through nozzles just beneath its mandibles on either side of its mouth. When the time comes for it to pupate, it begins to wave its head from side to side, extruding the threads like a conjuror pulling ribbons out of his mouth. They are slightly sticky and soon the larva has created a loosely textured wrapping around itself. It continues working until it has built this up into a dense fuzzy capsule about the size of a walnut, some white some a bright yellow. Ten thousand years ago in China, human beings discovered how to unwind these filaments, spin them together into thicker threads and then weave them into a fabric. In spite of our modern mastery of the techniques of chemical synthesis, we have not as yet succeeded in exactly replicating the molecules of

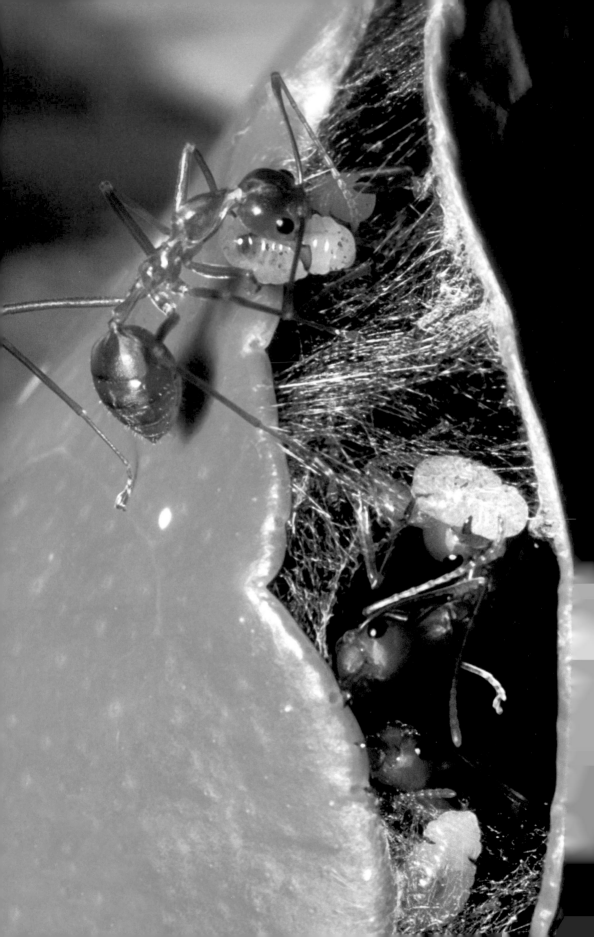

natural silk. So the most sumptuous of all our fabrics are still those that are woven from filaments produced for us by caterpillars.

Ants, like us, weave with silk produced by larvae. In their case, however, the larvae are their own. Green tree ants construct their nests from the living leaves of trees. A platoon of workers will march down a stem and along the edge of a leaf. Gripping with the tiny claws on their legs, they lean over to a neighbouring leaf, seize its margin with their jaws and draw the two leaves together.

Another squad of workers follows behind, each carrying a tiny white larva in its jaws. They approach the junction of the two leaves from the underside. Once in position, a worker gives the larva between its jaws a small but visible pinch and in response the larva obediently starts to extrude silk threads from the glands on the tip of its abdomen. While the ants holding the leaves together maintain their grip, the workers pass their living tubes of glue back and forth across the junction and quickly build up a white silken tissue between the two leaves. They repeat the process with other leaves and before long they have constructed a green ball the size of a grapefruit within which the colony will live.

Other larval insects use silk to trap their food. Descend into one of the limestone caverns around the little town of Waitomo in New Zealand's North Island and your breath will be taken away by a light display of astonishing beauty. The ceiling of the cave is covered by a galaxy of tiny lights. They shine from every wrinkle, curtain and pendant of stalactite. Below they are reflected in the black surface of the river which still flows through the cavern it created.

The marvellous lights above you are often compared to the stars and certainly they have something of the glory of the Milky Way that shines outside. But these subterranean lights are not coldly white. They are a wonderful electric blue. They are also very evenly spaced and they do not twinkle but shine with an unvarying steadiness. A more apt comparison might be to Christmas illuminations – were there any so restrained in the range of their colour yet so lavish in their number.

◁

Adult green tree ants cannot themselves produce silk, but their larvae can and have specially large silk glands beside their mouths that enable them to do so. The adults therefore are able to use their larva like tubes of glue to join leaves together when nest-making.

▷▷

The alluring, glowing bodies of fungus gnat larvae illuminate the ceiling of a cave in Waitomo, New Zealand.

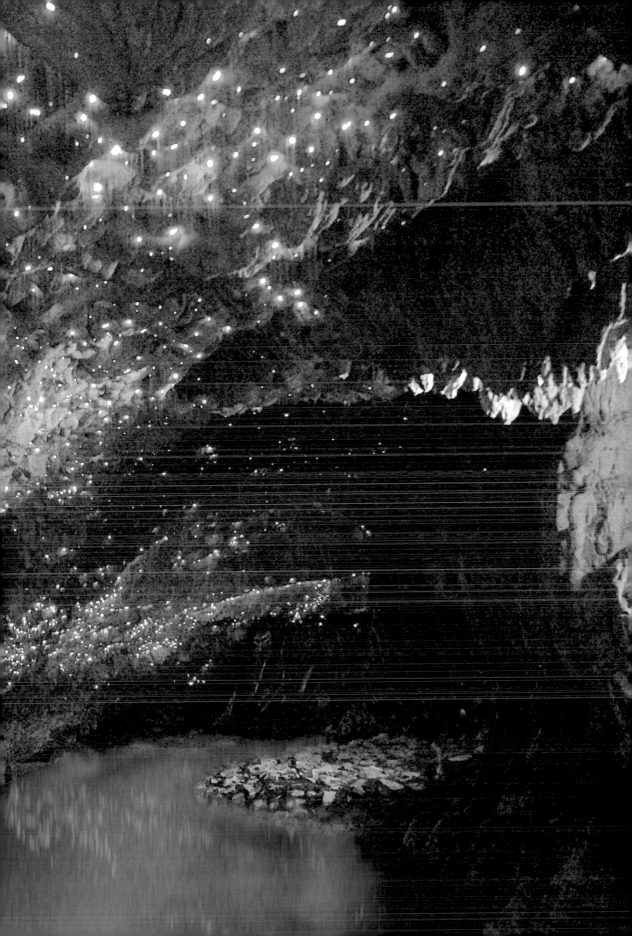

If you shine your torch obliquely across the ceiling, you will see the silk. Each pin-point of light is surrounded by curtains of fine dangling threads, each several inches long and each loaded with glistening evenly-spaced beads of glue like strings of pearls. On the ceiling in the centre of each cluster, in a horizontal tube of mucus and slung by silken threads from the ceiling, lies the insect that is producing the light. It is the larva of a fungus gnat.

Most of the members of the fungus gnat family are vegetarians. Many, as their name implies, feed on the fruiting bodies of fungi and are a pest as far as human mushroom farmers are concerned. The Waitomo fungus gnats, however, differ from most of the family – they are carnivores.

In the bush outside, they live in humid shady places, such as river banks or beneath broad leaves in the undergrowth, where there is little wind to disturb and entangle their hanging threads. But the conditions in the cave suit them even better, for here there is no wind at all, the air is always humid and it is permanently dark so that their shining lures are continuously visible no matter how bright the sunshine in the world outside.

The river brings down the larvae of midges that were laid in the mud farther upstream in the outside world. Here in the cavern they hatch, and as the adults rise to the surface of the river they glimpse above them the lure of the fungus gnats' lights. Up they fly, only to be caught by the sticky beads on the silken lines.

A gnat larva, lying in its mucus hammock, quickly detects the vibrations of the struggles of an entangled fly. It wriggles half-way out of its tube and closes its tiny jaws on the silk line a little distance down from its fastening. Holding on to its tube with the rear half of its body, it contracts its front half so hauling up its catch. Then it reaches down, seizes the thread with its jaws and starts to eat it. The trapped insect is hauled higher and higher towards the ceiling until it reaches the larva's jaws and there it is consumed. The silk filament with which the fungus gnat larva works is a double thread but even so it is extremely thin. Yet it has such great tensile strength that it can support a weight of 15 milligrams.

◇

A fungus gnat larva lies in its mucus hammock with its rear end glowing, while it waits for insects to blunder into the dangling silken threads laden with drops of glue with which it has surrounded itself.

116

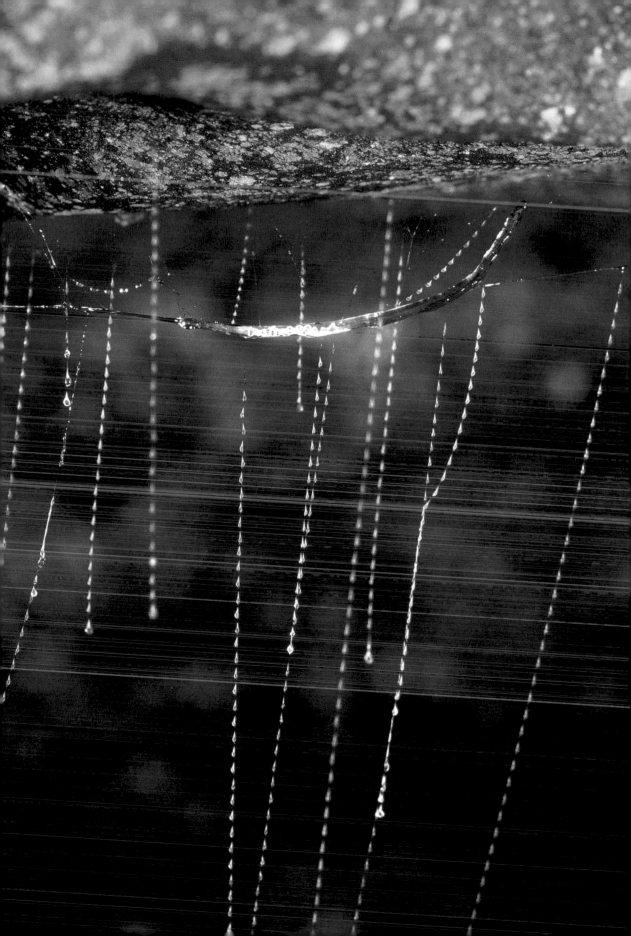

Very few adult insects produce silk. Lacewings are among them. They use it to protect their eggs and in a most ingenious way. When a female lays, the first thing to emerge from her abdomen is a tiny silken blob which she dabs repeatedly on the twig on which she stands until it sticks. Then she lifts up her abdomen and as she does so a hair-thin stalk of silk is drawn out from her. Under the tension, the emerging silk solidifies into a rigid bristle that lengthens as she raises her abdomen still higher until finally an egg emerges on the top of it. Its supporting silken stalk is so thin that insect predators such as ants may run along the twig within millimetres of the eggs without realizing that a tasty and nutritious meal could have been collected from immediately above their heads.

Another insect family, the embiopterans or web-spinners use silk in a particularly extravagant way. The adults have a pair of small feelers on the end of their abdomen that make them look a little like earwigs though they are more closely related to stick

△
Lacewing eggs, each suspended out of harm's way on a slender stalk of silk attached to the underside of a leaf.

118

insects. They produce their silk, uniquely, from their forelegs, squirting it from a hundred or so hair-like nozzles on the legs' undersides. As they wave their forelegs from side to side, with an action not unlike that of someone plastering a wall, the silk emerges in broad glistening sheets. With this they carpet considerable areas of the forest floor and swathe tree trunks up to heights of 3 metres or more.

They live beneath these sheets, grazing on spores, lichens and algae, safe from birds and other predators. Although the sheets are so thin as to be almost translucent, they are nonetheless waterproof so that the air within the tents remains moist. When one of the inhabitants is thirsty, it doesn't venture outside. It simply locates a drop of water lying in a dimple on the sheet above, bites a tiny hole through the silk and takes a drink.

Web-spinners tend to run backwards almost as frequently as they do forwards and with equal enthusiasm. The females are wingless so this presents no problems. The males, however, are fully winged, but they have a mechanical safeguard to prevent

Web-spinners produce silk so liberally from multiple spigots on their forelegs that it forms sheets rather than separate threads, and with it line tunnels and erect tents.
▽

their wings catching on their low silken ceilings when they go into reverse. Each wing has a crease across its middle and will flip back if it snags.

Eventually a web–spinner colony grows so large that the young leave it and try to find new feeding grounds for themselves. The wingless females venture out from their cover and start to explore, running over the forest floor and up tree trunks, systematically biting whatever lies beneath them to discover, apparently, whether it is the sort of surface on which lichen and algae are likely to flourish and therefore a suitable site for a new encampment. The males also emerge. They travel farther to search for mates from different colonies. To do that they need to fly. And their wings do not let them down, for the males are able to pump extra liquid along the veins in order to make them rigid and so capable of supporting them in the air.

△
The branching silk-lined tunnels of a colony of web-spinners swathing the trunk of a tree, Trinidad.

The extent to which silk production is found among land invertebrates today suggests that it was also widespread among the very earliest terrestrial pioneers. But one group of their descendants has elaborated and exploited that ability to a greater degree than any other – the spiders. Among insects, only a few families produce silk. In contrast, all spiders do so. And whereas any one species of insect only produces one kind of silk, some species of spider can produce at least eight.

The spiders' ancestry is very different from that of the insects. Their origin lies in the group that gave rise to the horseshoe crabs, the amblypygids and the scorpions. In consequence, their anatomy is fundamentally unlike that of insects. They have eight legs not six like an insect and their bodies are divided into two not three parts.

Their front half is aptly known as the cephalothorax, for it represents the head and thorax of their remote ancestors that have become fused into one. Like an insect's head, it has at the front a mouth which here is flanked on either side by a pair of poison fangs. It also carries the primary sense organs. And like the thorax of insects, it carries beneath it the legs – in this case, four pairs of them.

Spiders can have up to eight pairs of eyes, usually placed along the front edge of the cephalothorax but also, in some species, a little further back. Some are tiny specks capable of registering little more than the level of light. Others are much larger – lustrous brown beads which shine in the light of your torch because they have a reflecting layer at the back, as cats' eyes have. Nonetheless, even these big eyes are relatively simple in structure and cannot produce detailed images comparable to those provided by the compound eyes of insects. Some spiders, in fact, are virtually blind. All of them rely on their sensitivity to vibrations as the primary source of information about what is happening around them.

In the place where one might expect antennae on an insect's head, and looking not unlike them, a spider has palps. These are packed with sense organs which detect not only vibrations but scent. Those of a male are always much bigger than those of his female, for he has an additional use for them. He copulates with them.

The cephalothorax is connected to the spider's back half, its abdomen, by the thinnest of waists. Through this short link run the nerve cord, the main blood vessel and the gut. The abdomen may be globular or elongated. In some species, it is camouflaged. In others it is spectacularly patterned and coloured or ornamented with spines. It contains the bulk of the spider's digestive tract, its long tubular heart, its genital organs, two pairs of lungs and its silk glands.

These glands fill much of the rear end of the abdomen. They open to the exterior by spinnerets which look rather like a pair of tiny fingers. A sudden increase in blood pressure forces liquid silk out of the nozzles on the spinnerets. As it emerges, the spider pulls it with a leg, so causing the chains of silk molecules to bond together and solidify into a filament. Spider silk is very thin indeed – about 0.0003 millimetre, only a tenth of the diameter of the threads produced by the silkworm moth.

There are several pairs of silk glands in a spider's abdomen, each producing a different kind of silk. The nozzles are muscular and can physically change the diameter, strength and elasticity of the thread that issues from them. They can give the threads complex internal structures, folding them up upon themselves or wrapping them in spirals around other filaments. So an individual spider can produce several different silken threads, each with different properties to suit particular purposes.

One is exceptionally strong, relatively unstretchable, and used for the spokes of webs. Another is elastic and coated with glue for catching prey. Yet another is used for wrapping the prey, once caught. And all female spiders have one more kind of gland than their partners, for they need to have a special silk with which to wrap their eggs.

◇

It seems very likely that spiders, right from their earliest beginnings, used silk to make homes for themselves in much the same

▷ *A North American orb-web builder,* Argiope, *crouching on her untidy web, illustrates the basic features of spider anatomy – eight legs, and a two-part body, the cephalothorax at the front separated by a waist from the fat abdomen.*

way as web-spinners do. The most primitive of living spiders is *Liphistius*, a genus that is found in mainland southeast Asia from Burma to the Malaysian peninsula. It is quite a large creature, as spiders go, measuring about 6 centimetres (2 inches) in length and brick red or a spectacular glossy black in colour. It still retains clear relics of its segmented ancestry for its abdomen, uniquely among spiders, retains a line of plates – between six and eleven depending on the species – that run down from its waist to the tip of its abdomen on both its upper and lower sides. There are also vestiges of segmentation in its internal anatomy – in the bunching of its muscles and the constrictions in the long tubular blood vessel that constitutes its heart. Looking at *Liphistius*, the connection between spiders and scorpions is clear, and indeed the oldest spider fossil, a fragment bearing a spinneret that was found in 380 million year old shales near New York, probably belonged to a member of this group.

△
One of the most primitive of spiders, Liphistius, *has an abdomen that on its surface retains visible traces of segmentation, evidence of the connection between spiders and scorpions. Its pair of palps are greatly enlarged so that they are almost as long as its legs.*

Liphistius lives in holes, often in a bank, which it lines with silk. It covers the entrance with a limp wafer-like flap made from fragments of moss, dead leaves or lichen glued together with silk which matches its surroundings so closely that it would be difficult to spot, were it not for half a dozen or so silken strands on the ground which radiate around it like the lines on a sun dial. These run beneath the flap to a silken collar that encircles the mouth of the hole. If you gently lift the flap with a twig, which is quite easy to do, you are likely to see *Liphistius* crouched immediately beneath, with six of its eight legs held together in front of it. Its back pair are outstretched behind it, keeping a hold on the tunnel's silken lining.

Silken threads radiating around the door of a Liphistius *burrow will tell the owner, waiting within, of any prey that might wander nearby.*
▽

It spends much of the night there. If a cricket or some other small creature crawling over the ground happens to touch one of the strands, *Liphistius* immediately dashes out, and within a fraction of a second has grabbed its prey and returned to invisibility behind its door. The speed of the action is so fast that you can barely follow it with the eye; and the accuracy of the pounce is so

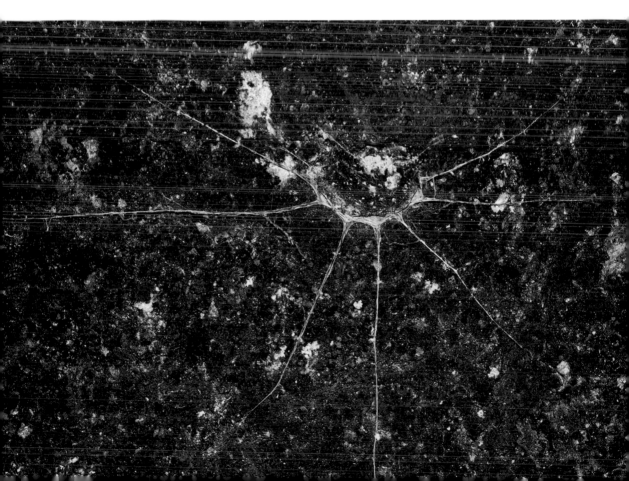

precise that you can have little doubt that *Liphistius* knew the exact position of its prey even before it emerged. It deduced it from the vibrations of the silken strands which it was touching with the tips of its legs.

◇

Liphistius has close relatives that have lost all signs of segmentation from their abdomens and are therefore considered to be more advanced in evolutionary terms. They are known collectively as mygalomorph spiders. They are large, mostly hairy and have a pair of formidable claw-shaped poison fangs in front of the mouth. If you approach one, it will threaten you by lifting up its forelegs and exposing these fangs which, in some species, may literally drip with venom. If you persist, the spider will strike. It slashes down-wards with its fangs which, if they connect, will open up a cut and at the same time inject poison.

One group of mygalomorphs, the true trapdoor spiders, have jaws equipped along the cutting edge with a row of spikes like a

A Western Australian mygalomorph spider, threatens with its huge poison fangs exposed and ready for action.
▽

A true trapdoor spider opens its door and pounces before a wandering cockroach has time to flee. Japan.

rake. With these, they dig a burrow that descends vertically into the earth. They waterproof its walls by plastering them with a mud made from earth mixed with their saliva, and line them with silk, working head down from the bottom of the tunnel upwards. When they reach the top of the tunnel, they continue weaving the silken sheet so that it closes the tunnel entrance like the skin on the end of a sausage. Next they cut around half its circumference to produce a neat flap. Then they emerge and add alternate layers of silk and mud to the top of the flap so that it forms a firm lid of exactly the right size with the uncut portion of the silk serving as a hinge. Some species even bevel the edge so that the lid fits precisely flush to the surface of the ground.

The fit is usually so perfect that it is even more difficult to find these tunnels than it is to detect the door-flaps of *Liphistius*. Your best chance of finding one is at night, for then the spider, crouching at the top of its shaft, lifts the lid a fraction of an inch and pokes out its two front legs, ready to launch an attack. During the day, the door is kept tightly shut and is so well camouflaged it is

127

virtually invisible. Some species, when digging, leave little piles of the soil produced by their excavations quite close to the entrance and these can be an indication that a closed door is nearby. One Australian species builds low circular walls of twigs and pebbles to prevent rainwater from flooding its home and these too may catch your eye. If you do, at last, find one, opening it is not easy either. The best way to do so is to insert a knife blade beneath the rim of the door. But it does not immediately lift up. The spider inside, alarmed by the vibrations in the ground that you will almost inevitably have created, has sunk its fangs in the underside of the door and braced its eight legs around the walls of the tunnel. Experimenters have discovered that the spider can withstand the pull of 38 times its own weight. However, it cannot maintain this grip for very long. Eventually it tires and retreats to the bottom of the tunnel so you can lift the lid and peer inside.

The tunnel may be a simple one that descends vertically several inches. Some trapdoor species, however, elaborate their security arrangements. One adds a small chamber half-way down the vertical shaft that opens horizontally to one side. The spider spends most of its time in this side chamber and fits it with its own silken door. Another species builds a similar side chamber but instead of living there, puts a specially moulded pear-shaped pellet of mud inside it. When this pellet is standing vertically, it almost fills the side chamber. But attached to its top edge is a sheet of silk that extends downwards to form the lining of the main vertical shaft. If danger threatens, the spider quickly descends the shaft and pulls down the mud pellet so that it hinges on the lower rim of the side chamber and falls over into a horizontal position. Its thinner half now effectively blocks the shaft, but the thicker end is still within the side chamber. An intruder will therefore conclude that the shaft is not very deep and empty. When the spider considers that danger has passed, a push from beneath sends the pellet back into its vertical position, like a teetotum doll righting itself, and so opens the way to the surface once more.

Trapdoor spiders are sizeable creatures – up to 3 centimetres (an inch or so) long – but other mygalomorphs are giants.

△
Theraposa *one of
largest so-called
bird-eating spiders
from French Guiana
with a captured
mouse.*

Theraphosa, a hairy monster from South America is the biggest of all living spiders and has a leg span of 26 centimetres (over 10 inches). These alarming creatures live in dens beneath a leaf or a stone which they line with silk. Some species lay trip-lines as *Liphistius* does. They extend threads from their den lining and stretch them across the ground in front of their thresholds. They are commonly called bird-eating spiders but if they do in fact catch birds, that is very exceptional. They probably acquired their daunting name because one of the books that first brought these creatures to the attention of Europe in the eighteenth century had a sensational illustration showing a spider sitting beside a dead bird. That was a reasonable way of indicating the spider's great size but, intentionally or not, it gave the impression that the spider was standing victoriously beside a typical victim. Their more usual prey is, in fact, insects and very occasionally, small rodents.

There is a third group of mygalomorphs – the funnel-webs. They are not as big as the bird-eaters or the trapdoors but they are

129

among the most venomous. The most notorious is the much-feared Australian species, the Sydney funnel-web, *Atrax robustus*. It usually builds its silken tube in piles of twigs or other dead vegetation and surrounds its entrance with extensive silk sheets. Any creature that strays on to these sheets will be grabbed and killed. And if a human being interferes with the funnel-web, it will defend itself with a bite. The effects of its venom are fearsome indeed – extreme pain, unbearable cramps, vomiting and finally convulsions that may, in some cases, lead to death.

All the rest of the spiders in the world are more specialised and highly evolved than the mygalomorph spiders. Their fangs, with the exception of one small and strange family, operate in a different way. Instead of slashing downwards, they work horizontally, more like a pair of pincers. And instead of breathing by means of two pairs of book lungs, similar in principle to those of scorpions as bird-eating spiders do, at least one pair of their lungs has been replaced by simple branched tubes that carry air directly to all the organs of the body, similar to the tracheae evolved by insects.

There are twenty-six different families of these more advanced spiders and they contain the vast majority of the 36,000 or so different species of spider that exist today. Between them, they have evolved a great variety of life styles and hunting techniques.

Crab spiders are so called because they tend to move in a scuttling sideways fashion. Like all spiders they continually produce a single filament of silk that trails behind them. These drag-lines serve as trails that inform others of an individual's presence and identity. More importantly, they are also means of escape in a crisis. If alarmed, a spider can drop down on a thread and when danger passes ascend by simply eating the silk from which it hangs. Crab spiders, however, do not use silk to construct any kind of home for they have none. They spend their time crouched in the vegetation waiting for a victim to come by. Some lurk on flowers, and are excellently camouflaged and able to some extent to change their colour to match that of the petals on which they crouch so they are not always easy to see. They rely on the

▷
A crab spider lurks on a flower, its abdomen so closely matching the pink of the petals that it is almost invisible either to our eyes or to those of an insect that might be about to visit the flower in search of nectar.

flower's colour, perfume and nectar to lure butterflies, bees and other insects within their reach.

Wolf spiders are wanderers. They range widely across the ground seeking prey but they also have burrows into which they retire when they are not hunting. This family includes particularly large species which – although they cannot rival bird-eating spiders – nonetheless achieve a length of 4 centimetres (1½ inches). One of these, which builds a silk-lined tube for itself as a home and sometimes even equips it with a door, is known as the tarantula.

◇

There is some confusion about the name tarantula. It was originally given to this species because it is common around the southern Italian town of Taranto. It is a big spider by European standards, with a body than can be 30 millimetres (1 inch) long, and it is greatly feared. It was believed during the Middle Ages that the only cure for anyone bitten by such a spider was to dance violently and continuously until the patient fell down from sheer exhaustion. So a very vigorous dance popular in the region then acquired the name of tarantella. Indeed groups of musicians traditionally travelled throughout the countryside ready at all times to assist in a cure.

The tarantula is one of the biggest of European spiders, so when big hairy bird-eating spiders were discovered in the New World, even though they grow to a much greater size than the original tarantula, they too were given that name. Now 'tarantula' is widely used on both sides of the Atlantic as a name for the mygalomorph bird-eating spiders of the New World.

But there is further confusion. When scientists came to give a technical name to the original Italian spider, they called it, in recognition of local tradition, *Lycosa tarantula*. But this species does not in fact have a very poisonous bite. It is said to be little worse than a wasp sting. The true culprit, for which dancing was thought to be the best cure, was probably not the big conspicuous hairy monster but a rather less impressive species, *Latrodectus*, a

◁
A wolf spider hunts primarily by sight and accordingly has a pair of very large eyes as well as six other smaller ones.

133

black spider about the size of a thumb-nail with red spots on its abdomen. That too is found around Taranto as well as many other parts of southern Europe – and its bite is very dangerous indeed.

Latrodectus belongs to a family of spiders that are sometimes called the comb-footed spiders because of the tiny bristles on the ends of their feet which enable them to cling to the smoothest surfaces. It is widely spread throughout the warmer parts of the world and is so feared that everywhere it is given a special name. In France it is known as the malmignette; in Australia the red-back; in New Zealand the katipo; in Mexico the araña capulina; in Chile the araña del lino; and in the southern United States, the black widow. Within an hour of being bitten by any one of these, the victim is in agony. Severe pain and muscle spasms gradually spread through the body. If they reach the chest and the breathing muscles, they may cause death so it may well be that the traditional dancing of vigorous tarantella might have some therapeutic affect by raising the body temperature and dispersing and therefore diffusing the venom's poisonous effects. Happily today there is an anti-venine available and deaths have been reduced to less than 1 percent of those bitten.

Members of the family known as jumping spiders do indeed

This extremely poisonous southern black widow Lactrodectus *spider in Georgia, USA, has her egg sac beside her on her web.*
▽

jump and spectacularly. Some can cover forty times their own
length in a single leap. To do that they need good eyesight and so,
not unexpectedly, they have the best of any spider family – eight
eyes in three rows. One pair is very large and occupies most of the
front edge of the cephalothorax. Each of these eyes has muscles at
the back which can move the eye inside the head. So jumping spiders can look from side to side without moving their bodies at all.
Having spotted their prey from several inches away, they use their
two back pairs of legs to spring upwards so that they sail through
the air with their front legs reaching forward ready to grasp their
prey or to give them a firm landing.

Spitting spiders catch their prey in a way that is all their own.
Scytodes, a British species, is a slow-moving nocturnal creature. Having sighted its victim it squirts gum from its fangs shaking its head as it
does so, with the result that the gum forms a zig-zag line of bonds
that binds its victim to whatever it is sitting on. With it thus immobi-
lised, *Scytodes* cautiously advances and delivers a killing bite.

These spider families became established very early in the

135

group's history. They doubtless took a heavy toll of the ancient flightless insect ancestors such as silverfish and bristletails. It may well be that their very success as predators was one of the factors that gave impetus to the insects' development of flight. Certainly the move into the air around 350 million years ago must have taken much of the early spiders' potential food out of their reach. But not for long. Spiders used their silk to trap prey in flight.

One of the simplest techniques was to stretch threads in all directions between tall plant stems or bush twigs and the ground. *Linyphia*, the domed-web spider, constructs sheets of silk beneath a bush. Above, in the twigs, she hangs threads running in all directions. An insect, even quite a large one, flying beneath the bush is almost certain to crash into one of these threads. Its speed and weight may be sufficient to break the first thread it strikes and it may continue onwards even though it has been slowed down. Eventually however, as it blunders from one thread to another, it is brought to a halt and is so unbalanced that it falls on to the sheet below. *Linyphia* is ready and waiting. She runs along the underside

The simple cone-shaped but somewhat untidy trap built in an English meadow by Linyphia, *the domed-web spider.* ▽

of the sheet, quickly bites her catch and drags it through the silk to bind it up and hang it in her larder.

One family of sheet-web makers has elaborated the technique. They build funnels that lead away from the main sheet and here they, in safety, await their victims. The house-spider, *Tegenaria*, is one of these. But one species has found the relative warmth and dryness of human dwellings so much to its taste that it has moved in with us wherever we live in the world, and is now found nowhere else. They are the ones that spin the cobwebs that appear in undusted corners and live in a tube that leads from them. With these webs they trap cockroaches and flies and to that extent are to be welcomed rather than the reverse. The one we sometimes find running around helplessly in our empty baths, desperately and unsuccessfully trying to get a grip with its feet on the smooth enamel is almost certainly a *Tegenaria*, often a big male that has left its home in search of a mate. It is true that it may bite if you pick it up, but the effect is scarcely as damaging or as painful as a slight pin-prick.

One spider is so extraordinary that it has been given a family all

A sheet-web builder, Agelena, *adds a funnel to her web so that she can lurk nearby.*
▽

of its own. There is only one species and it lives all over Europe and Asia. It looks not unlike *Tegenaria* and was once classified with it. There is nothing about its anatomy which might give you a clue about its particular and extraordinary life. Indeed it is a tribute to the enterprise of spiders that this one, *Argyroneta*, has managed to create for itself a unique life-style with such evolutionary speed that it has not yet had time to develop any particular anatomical adaptations to aid it in its new life. Its adaptations are entirely behavioural. It lives underwater.

Like any other spider, it breathes air and comes to the surface to do so. When it dives down again, it takes with it a bubble caught by the hairs of its abdomen so that it appears to be invested in silver. But it does not merely dive occasionally. It makes a permanent home underwater.

In describing web-spinning spiders at work, it has become customary to refer to them all as female. There is some justice in this, for although some male spiders do spin webs, the biggest produced by the species are most usually the work of the females. And spiders when you watch them, seem to have such complex and individual personalities that it seems preferable to call one 'she' rather than 'it', even though the one being described may, sometimes, be male.

Argyroneta begins construction by swimming down a few inches below the surface of a pond and making her way into a clump of water plants. There she starts to clamber between the stems, trailing a line of silk behind her. She has to interrupt her labours every now and then to return to the surface and renew the air around her abdomen. Soon she has created a relatively dense tangle of threads. When she is satisfied with it, she swims up to the surface yet again. This time she captures a large bubble of air between her two long hind legs. Using her other six legs to swim, she laboriously makes her way downwards, dragging the bubble behind her. The buoyancy of the bubble makes this very hard work and since *Argyroneta's* legs are not flattened in any way she has to flail them very vigorously to make any head-way through the water. Eventually she manages to descend to a point directly

▷

A water spider, Argyroneta, *adds another air bubble held between her hind legs to the silver chamber she has already built among the rootlets at the bottom of a pond.*

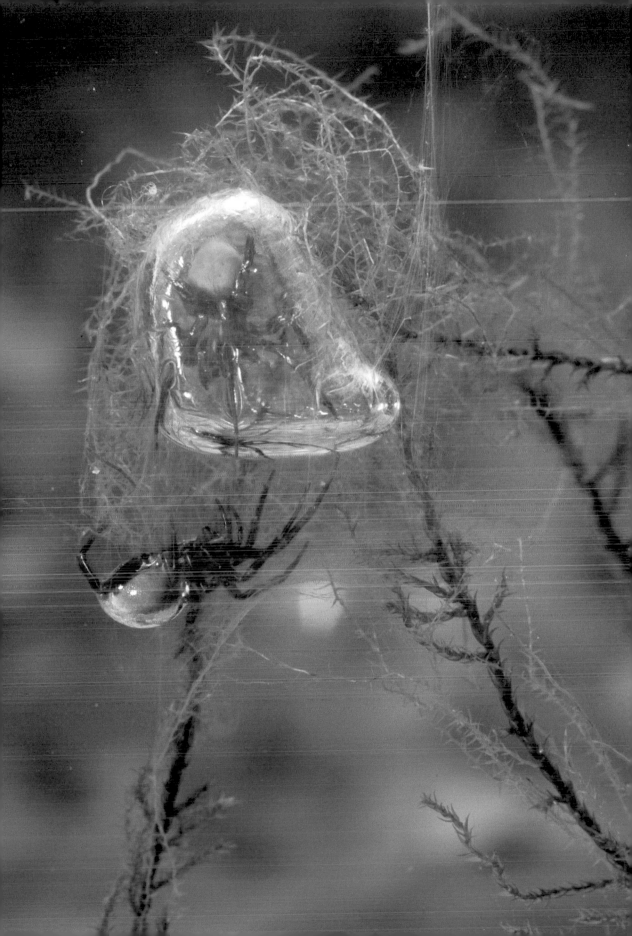

beneath the silken tangle. There she releases the bubble which rises up and is trapped by the silk threads. She also wipes away any tiny bubbles that have been caught in the hairs that cover her abdomen and they too rise, increasing the size of the main bubble. The water spider now has the basis of her underwater home.

She returns again and again to the surface to bring down more bubbles. The silk of the threads *Argyroneta* has used is an elastic kind and it stretches as the bubble beneath the threads grows. Soon the spider has to add more threads to prevent the bubble breaking loose. Eventually she has an underwater air chamber the size of an acorn enclosed by a bell-shaped membrane of silk. Now *Argyroneta* can stay underwater for long periods, breathing the air in her bubble and waiting for prey.

The water spider is not large – about half an inch (12mm) – but she will tackle insects of all kinds which she seems to detect from the vibrations they make in the water. Many are land-living insects that have fallen accidentally into the water and lie struggling on the surface. *Argyroneta* swims up to grab them but has to ferry them down to its bell-chamber before she can eat them for like all spiders, she feeds by saturating her prey with digestive fluids from her salivary glands while it is still in her jaws and that can only be done in air.

The spectacular webs we see in our gardens, slung between improbably distant anchor points, hung with dew on an autumn morning or occasionally and most wonderfully, covered in frost, are the work of orb-web spiders. *Araneus* is one of the most common.

Work begins at night, usually an hour or so before dawn. A female climbs up to a prominent perch. She may have used it many times before and knows it to be suitable. There she pauses, head down, abdomen upright, and begins to spin a line, the end of which she attaches to her perch. If the night air is totally still, she crawls down and across and then up to another anchor point trailing the silken line behind her which she hauls in until it is taut. But there is nearly always a slight breeze which catches the

▷
Araneus, *the common garden spider, at her daily labour of constructing a web. England.*

specially thin filament issuing from her spinneret and carries it away horizontally. With luck and more often than not if this is a stance she has used before, the filament will catch on another anchor point some distance away – a twig, a leaf, a window frame. It may be a metre or more distant and if the thread has been carried by the wind, it may span a path or even a stream. Now she has the basis for her construction work. She hauls the thread tight, fastens it to her perch and climbs along it, rather gingerly, for this filament, the better to be airborne, is very thin indeed. But she trails behind her a thicker thread. This has to be particularly strong, for it will have to carry the whole of her finished construction. Sometimes, as she goes, she eats the first flimsy line. Silk is precious and is not to be wasted. While she is doing that she herself forms the link between the first line and its stouter replacement.

Once across, she continues to spin her second heavy-duty cable. This, because of its weight, droops slightly. She secures it by

△
Araneus, *halfway through her construction work. Having established the radiating spokes of her web, she is now adding a sticky capture line, spiralling inwards from the outer margin towards the centre.*

142

pressing the sticky end against the anchor point. Back she goes to the centre of the cable, sticks another line from her spinnerets to it and lets herself down into midair, hanging from the silk thread that issues from her spinneret until at last, drifting perhaps in the faint breeze, she makes contact with a lower anchorage. Once there, firmly gripping her stance with her legs, she pulls the thread tight. Now she has a Y-shaped basic framework. The meeting point of the two arms of the Y will become the centre of her web. She then proceeds to add as many as eighty spokes, radiating from this centre point.

As she busies herself with this work she also leaves a number of threads connecting the spokes at the very centre. These she now surrounds with two or three circles to give the central hub special strength.

Her next task is to lay a line that starts a little farther out from the central hub and spirals across the spokes to the outer margin. Like the initial horizontal line, this is only temporary. Having reached the outer margin of what will be the completed web, she returns, spiralling back to the centre, eating the guide line as she goes and laying in its place a sticky capture line. She stops short of the hub so that there is a space, a free zone, that allows her when necessary to dodge through the spokes from one side of the web to the other.

This sticky thread is very remarkable. It is perhaps the most complex of all the kinds of silk produced by spiders. It has to have properties very different from those of the silk forming the spokes. They must be strong, rigid and relatively non-elastic so that the web retains its orb shape even when blown by the wind or put under strain by the struggles of prey. But the capture spiral is different. It must be sticky enough to hold the prey and it must not break. Such strength is not necessary for the silk with which ground-dwelling spiders build their traps. If one of their filaments breaks, others will entangle the prey. But in a two-dimensional orb-web, the breakage of a single thread may allow a catch to escape. The capture line must therefore be elastic enough to stretch when a flying insect cannons into it. It must also be

resilient enough to return to its original length almost immediately, for if it did not, it would sag and risk sticking to other parts
of the spiral, thus distorting the web as a whole or even tearing it.
On the other hand, if it were to spring back like a rubber band or
a trampoline, the insect might well be catapulted off again and be
lost. So the capture line must return to its original length with virtually no recoil.

This specification seems nearly impossible. The capture line
fulfils it, not by its material substance but by its microscopic structure. The filament that issues from the spinnerets consists of a pair
of fibres. As they emerge, they are given a coat of glue. This
absorbs water from the atmosphere which gives the silk an elasticity. As each length is completed and fastened to a spoke, the spider
twangs it with one of her hind legs. The vibration causes its liquid
covering to coagulate into a line of equally spaced beads. The surface tension of these minute globules pulls in the central fibres of
the filament like a windlass winding in rope. When an insect hits
the thread, these fibres are pulled out of the droplets. But when
the impetus of the insect is slowed to a halt, surface tension rapidly
rewinds the fibres back into the globules and the thread reverts to
its original tautness.

◇

Some webs have broad white ribbons stretched across their centre. They may form an X shape or a series of concentric rings.
Such a feature is called a stabilimentum but the term is a mislead-
ing one for there is no evidence that it does anything to improve
the stability of the web. Exactly what function it does have is still a
matter of debate. Some authorities believe that stabilimenta are
warnings to birds to discourage them from flying through the web
and destroying it. More recently it has been found that some
stabilimenta reflect ultra–violet light very efficiently and it has
been suggested that since many flowers have ultra–violet markings
on their petals to guide insects into them, the stabilimentum, far
from being a warning to deflect other creatures, is a deceptive lure
that entices insects into the web.

▷
*A Malaysian
orb-web builder
emphasises the
presence of her web
by habitually resting
at its centre with her
four pairs of legs
resting on the rays of
its stabilimentum.*

144

There is a danger in making the web conspicuous as the stabilimentum certainly does. Its owner becomes an obvious target for predators. But one stabilimentum spinning spider, *Argiope*, has a way of minimizing the risks. If she perceives a threat as she clings to the centre of her web, she braces her legs and with considerable muscular effort, vibrates the whole web so that it swings back and forth. That makes it very difficult indeed for a bird to focus upon it with sufficient accuracy to be able to pick the spider off the web without itself getting entangled.

Stabilimenta are constructed by spiders belonging to several different groups and must have evolved quite separately on several different occasions, so it is very possible that they serve different functions for different spiders. One group that constructs them uses a very special kind of silk. They produce it from a plate-like spinning organ that is placed just in front of their three pairs of spinnerets and is pierced by many thousands of tiny spigots. It is

△
A variable decoy spider, Cyclosa, *in the Thailand rain forest, sits at the centre of her spectacular stabilimentum.*

146

called a cribellum – the word is Latin for a little sieve. The spider draws silk from it by combing it with special hairs on its hind legs. Whereas normal threads of spider silk have a diameter of 0.001 to 0.004 millimetres, a single fibre of cribellate silk is 0.00001 mm thick. Many hundreds of these threads are spun together on support threads. The result, know as cribellate, fuzzy or woolly silk often has a bluish appearance when fresh. It is not sticky but traps prey in the same sort of way that fibres of wool can entangle something covered with hairs or bristles. It is so effective at doing so that members of one family of cribellate spiders that make orb-webs have no poison glands. They rely entirely on wrappings of cribellate silk to subdue their prey.

The whole process of spinning an orb-web may take an industrious spider no more than an hour. In that time she will have produced from her spinnerets some 30 metres (122 feet) of silk.

Many of these marvellous constructions are demolished by their builders every night. They may have been wrecked during the day by some large animal blundering through them and in any case the glue along their threads will have lost its stickiness by being covered with dust or drying out. But the spider does not waste their substance. She rolls up the silk and eats it so she has the where-withal to produce fresh silk the following morning.

It might be supposed that the actions needed to create these complicated structures are purely instinctive. And indeed young spiders right from hatching and without experiment or tuition, are able to spin small webs that are perfect in all their details. Yet it is also the case that if you snip some of the filaments of a web, its creator not only senses exactly where the trap has been damaged but skillfully and unhesitatingly repairs it. So a spider is certainly able to assess the construction of its web and take appropriate, individual action to improve it if necessary.

By dawn, an orb-web spider is ready and waiting for her first victim. Some species settle in the centre of the web, others lurk in the foliage at the edge, holding a long signal thread from the web by a front leg. If an insect blunders into the web, the owner dashes out. She knows exactly where to go from the differing vibrations on the various threads. Some species secrete oil on the tips of their legs which prevents them from sticking to the capture threads. Others run swiftly over the web skillfully placing their feet only on the unglued spokes.

A bite from poison fangs is enough to quell the victim. The spider then wraps it in silk, turning it round and round with her feet and either takes it back to the edge of the web to eat it immediately or hangs it in the larder beside the web.

The biggest of these orb webs are those made by *Nephila*, huge tropical spiders with legs that may span 20 centimetres (8 inches) and elongated black bodies brightly spotted with colour – sulphur yellow in some species, pink in others. The web-builders are all females, for the male is a diminutive creature a mere fraction of her size who shares the web with her. A *Nephila* web may be well over a metre across and the attachments of guy ropes supporting it six metres (20 feet) apart. It is frequently stretched across the open space of a track in the forest, where insects, small birds and bats regularly fly – the sort of space that you automatically choose as you make your own way through the forest. Blundering into one of these webs is, even for a human being, an unpleasant experience. The sticky threads are not easy to brush away from your eyes and mouth. They are, in fact, so strong and thick that in some parts of the Pacific, the local people pick them up on a frame and use them as fishing nets.

These formidable creatures are not, however, all-powerful, even in their own world. *Argyrodes*, a spider belonging to the same family as the black widow, is one of a number of species from several different families that have become pirates. It regularly targets *Nephila*. It builds a small rather untidy web near *Nephila's* huge construction and from it makes regular excursions on to the web of its giant neighbour. When *Nephila* makes a catch and starts to wrap her victim, *Argyrodes*, detecting what has happened from the vibrations of *Nephila's* web, moves slowly and stealthily towards the storage area of the web where *Nephila* hangs her wrapped meal. Very gently, avoiding vibrations as far as possible, *Argyrodes*

▷
One of the biggest of all web-builders, the female Nephila *lives in many parts of the tropics. This one has caught a lizard. The smaller spider clambering about on her catch is her male who lives on her web and claims a share of her catch. The small insect on the right is a biting midge, gorged with the lizard's blood.*

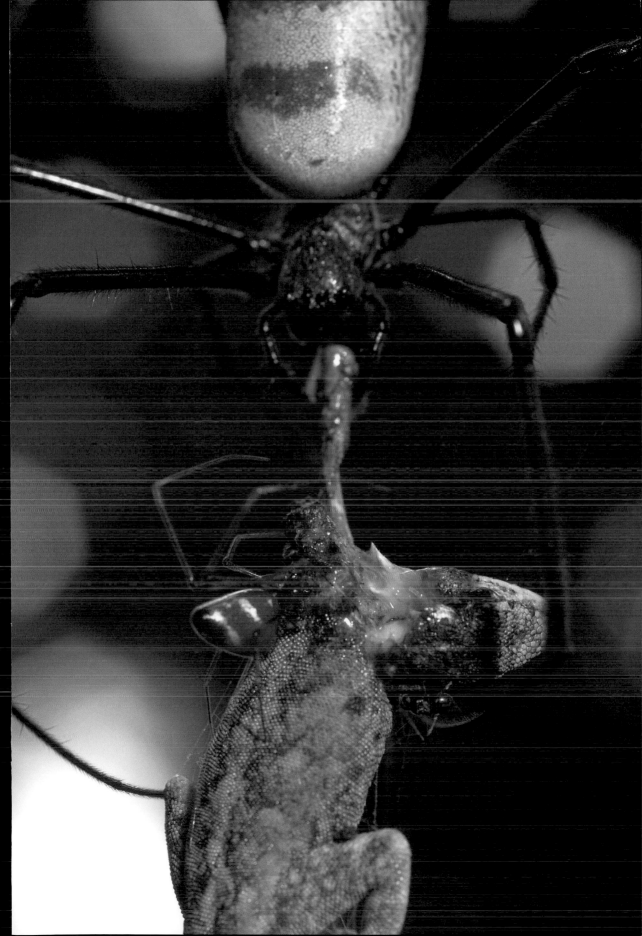

cuts the filaments that support the package until it is able to remove the meal. *Nephila's* webs are sometimes so big that many such pirates are able to make a living from it. Over forty have been counted on one single web. Nor are the pirates necessarily satisfied with stealing *Nephila's* catches. Sometimes they will attack the owner herself, subdue her with their venomous bite and make a meal of her

The most complex orb-webs are those spun by another tropical spider, *Uloborus*. It starts by building a normal orb, but places it not vertically but almost horizontally. It then runs a filament from the centre of the hub to an attachment, well outside the web, and hauls it tight so that the web is pulled into a cone shape. Then *Uloborus* builds another orb across the mouth of the cone and takes up a position in the middle of the two to await her victims.

Scoloderus, the ladder-web spider, has a different technique. A major part of many nocturnal spider's diet consists of moths. Some of these, however, have developed a defence against these entangling nets. The scales covering their wings are detachable so a moth flying into a web is able to free itself and fly away leaving behind nothing but a few wing scales. But *Scoloderus* has evolved a response even to this. She builds an orb web much like many others. But above it she constructs a series of horizontal threads, one above the other, like a ladder. If a moth with loose scales on its wings flies into the orb web, it may manage to free itself by leaving behind some wing scales. But when it strikes *Scoloderus'* ladder web it tumbles downwards. As it tries to continue its flight, it strikes more of the ladder's silken rungs until at last it has few wing scales left and becomes inextricably caught.

Some spiders use their precious silk less extravagantly than the orb-web builders. *Dinopis* does so in a way that has given it the popular name of gladiator spider. She first constructs a framework between the twigs of a bush using a dry silk. Clinging to this, she then builds a small square net using a sticky and very elastic silk. She chooses to do this in a place that is so secluded that she is very unlikely to trap passing insects. But having completed the net, she detaches it and holding it between her front two pairs of legs like a

▷
A gladiator spider, Dinopis, *hangs from her suspension line with her net held outstretched between four of her legs, waiting for unwary prey to walk beneath.*

150

kind of cat's cradle, she hangs head down from a filament she has spun beneath a leaf so that it is within a few inches of the ground. When an insect walks beneath her, she stretches out her legs, and quickly places it over the insect like a string bag and releases it. The highly elastic silk instantly contracts, and the insect is trapped in its threads.

Another spider, *Mastophora*, is even more frugal with its silk. She uses a device similar to the bolas used by Argentine gauchos to bring down cattle. That consists of a length of rope with a metal weight at each end, and another weighted piece tied to its middle. Skillfully thrown at a galloping cow, a bolas entangles the animal's legs and makes it trip and fall. The spider uses a bolas with just one weight, a blob of sticky liquid silk, and she keeps a hold on the other.

Bolas spiders are widely distributed around the world, though there are none in Europe. One of the North American species has a plump, bead-shaped body about the size of a pea and strikingly coloured in black and white so that it appears to be a bird dropping

△
Dinopis has pounced, trapping an insect in her silken net which, because it is made of elastic silk, has retracted and entangled her victim. Now she can eat.

that just happens to have fallen on a leaf. When evening comes, however, the bird dropping unfolds its legs and starts to spin.

First she lays a horizontal line of dry non-sticky silk along the underside of a twig or a leaf. Hanging from this with two of her legs she next produces another line an inch or so long, this time of a silk which is sticky. This is the one the free end of which she weights with a blob of silk. And there *Mastophora* dangles, with one long foreleg holding the hanging line. She may stay like that for a quarter of an hour or so. If she has no luck, she reels in the line and eats it, perhaps because her blob may have lost its stickiness. But within a few minutes she will try again. Cutworm moths are active in these first few hours of the night. If one approaches, then with a twitch of the spider's leg, she begins to whirl her bolas – and suddenly she has the moth in her jaws.

A bolas spider, Mastophora, is waiting motionless with her weighted silk filament hanging at the ready. At the sound of a moth's wings she will immediately whirl it.
▽

The cue that caused her to swing her bolas is not the sight of the moth, for in fact she is virtually blind and in any case it is dark. It is the sound of the moth's wings. That can be demonstrated if you record the faint sound and play it to the spider through a small earpiece held within half an inch of her. When she hears it, she immediately begins her bolas swing. The moth, for its part, seems almost wilfully to fly to its doom. That may, in fact, be the case, for the spider uses an invisible lure, a pheromone that is exactly the same chemically as the perfume used by her victim in its courtship.

After two or three hours, no more cutworm moths appear. So *Mastophora* pulls in her bolas, eats it and then rests for an hour or so. Around midnight, she spins a new bolas and prepares to swing it once more. This time the sound that stimulates her to do so has come from a different moth, the Smoky Tetanolita. But it too is snared by *Mastophora's* bolas. Astonishingly she has managed to change her pheromone, like a knowledgeable fisherman changing his bait to match the particular quarry that has started to become available.

◇

Courtship among all spiders is a complex business for the males have to deal with the same kind of problem as those other pioneering land-dwellers – millipedes and dragonflies. None of them have testes that are directly connected to the equipment necessary to introduce their sperm into the female's genital openings. The male dragonfly does so by using the rod beneath his abdomen. A millipede transfers his sperm with the special modified leg near the front of his body. A male spider uses his palps and – being a spider – silk.

Soon after finishing his final moult and reaching adulthood, he spins a small silken napkin. On to this he deposits a drop of sperm directly from a pore on the underside of his abdomen. Then he turns round, dabs his palps into the droplet and sucks up the sperm as if loading a hypodermic syringe. He is then ready to respond to any female he may encounter.

△
Spanish tarantulas,
Lycosa tarentula,
mating. Spain.

The males have a further problem. All spiders, without exception, are hunters. A male therefore has to ensure that any female
he approaches realises that he is a potential father for her young
and not a particularly obliging meal.

A male trapdoor spider is able to discover from any silken
thread he may encounter lying across the ground, whether the
individual it came from is a female and whether or not she belongs
to his species. Once he has found a thread of the right sort, he follows it to the trapdoor and knocks on it with his front legs. The
female opens the door and rears up at the sight of her visitor,
exposing her large fangs. The male responds by lifting his front
legs so that a spur on each of them catches on her fangs and
enables him to push her fangs outwards and sideways, out of
harm's way. Slowly he pushes her backwards, until he is able to
reach forward with his loaded palps and insert them one at a time
into the two genital openings on the underside of her abdomen.
This may take a minute or so. Once he has managed it, he disen-
gages and runs for his life.

155

His sperm, however, has not yet fertilised her eggs. The open-ings into which he inserted his palps are not directly connected to the female's ovaries. Each leads first to a small chamber, her exter-nal uterus. His sperm will remain here, fully viable, until such time as the female produces her eggs. They will then travel from her ovaries along a duct and into the external uterus, and there unite with the sperm.

Xytiscus is one of the commoner crab spiders that can be found in an English garden. A male, having spotted a likely mate, approaches her tentatively. When at last he is close to her, he grabs her by one of her fore-legs. She is likely to struggle, but he soothes her by stroking her with his legs. When she ceases to wrestle, he begins to crawl over her, back and forth, trailing a silk filament behind him. He fastens it down to the petal or whatever she is sit-ting upon until, before long, she is strapped down by a network of bonds. With her thus secured, he clambers round behind her, lifts her abdomen and crawls beneath her. Then at last, he inserts his palps, one at a time, into her genital pores.

A male crab spider, Xytiscus, *takes great precautions to avoid being bitten by his rather larger mate. He cautiously clambers around her body, gently swathing it in silken filaments.*
▽

This may take him as long as an hour to complete for it is not as simple a process as it might seem. Copulation is much more involved among crab spiders than it is with the bird-eaters and their relatives. The genital opening of a female crab spider, like that of most more advanced spiders, is protected by a curiously shaped plate, called the epigyne. This is crumpled in a complex way. The male's palp in many species is also contorted and has all kinds of bulges and spurs on it. But palp and epigyne match one another in the same way as a lock matches its key. So if, in spite of all the signalling, some male crab spider came to grips with a female belonging to another species, his work would be in vain for his key would not fit her lock. The system is also a great help for those scientists trying to identify a specimen, for the shape of the female's epigyne and the male's palp is a certain indicator of its species. Thus it is that technical works dealing with spider taxonomy consist largely of highly magnified drawings of their genitalia.

Female wolf spiders make small burrows for themselves which they line with silk. A male, like a male trapdoor spider, is able to discover from the smell of the silk whether or not the female is ready to copulate. That knowledge is important, for if he advances too close to a female who is not ready to mate or has already done so, she will eat him. Even if she is fertile and awaiting a partner, he has to proceed with caution and let her know before he gets within pouncing distance who he is and what his intentions are. He sends a sound signal by tapping his legs on the ground or on a fallen leaf. Some species have small corrugations on their palps with which they make a buzzing sound. Since wolf spiders hunt on the ground, they all have to have good eyesight and the males are therefore able to back up their mating calls with visual signals. The palps and forelegs of the male are usually conspicuously marked with black bands or coloured tufts of hairs and he waves these about with all the vigour of a naval signalman sending a message with flags by semaphore.

One European wolf spider, *Pisaura mirabilis*, has developed a mating ritual that has not yet been observed in any other species of spider. The male begins by catching a fly or some other insect. Instead of eating it, he carefully wraps it in silk and then, carrying it in his jaws, goes in search of a female. When he finds one he approaches with caution. As he gets within striking distance of her, he straightens his front legs so that his body tilts upwards and makes his gift very evident – and strategically placed. It is between him and her jaws. She bites into it. As soon as she does so, he relinquishes it and swiftly swivels his body around so that his head is downwards. While she is busy eating his gift, he ducks under her abdomen, swiftly finds her genital pore and inserts his palps.

The male partners of the orb-web-building females are particularly small. They do not take on any parental responsibilities. They do not maintain territories. Nor do they help in the raising of their young in any way. Their contribution to the survival of their species is limited to providing sperm, properly placed.

A female's web for a hopeful male however, can be a help not a hindrance. Web-spinning species, for the most part, have tiny eyes and very limited sight. Some are virtually blind. Their

▷
A male wolf spider, Pisaura mirabilis, *cautiously approaches a female, carrying a nuptial gift in his mandibles. She, however, is already gravid and takes no interest in him.*

perception of the world comes from interpreting the varying vibrations of their webs. Many females spend all their time crouched in the middle of their orb webs so a male has to cross that web in order to reach her. His signals therefore, can be vibratory ones. As he ventures on to the first filaments he starts to twang one of the strands with his forelegs producing a rhythmic and repeated vibration that signals his identity to the female. Even so, he ventures on to the web with visible trepidation and cautiously trails a silken safety line behind him as he goes. At the least sign that he is unwelcome, he will leap off the web and swiftly lower himself to the ground. He may have to make a number of attempts before finally he reaches his target.

When the male of *Argiope*, a European orb-web builder, reaches his female and starts to insert his palps, she gently wraps him up in silk as he works. Sometimes he is able to break free. But if this is his second mating, he often seems to lack the strength to do so and then she eats him.

A male of *Nephila*, the tropical giant web-builder, may only weigh a thousandth of his huge mate. He is even smaller than her normal prey and so he will venture on to her web without taking any precautions whatever. But he can usually mate with her without her even noticing.

◇

No female spider lays her eggs immediately after copulation. Bird-eating spiders may wait several months before they do so and most others will take at least a week. But then, once more, silk becomes essential. In most species, the female starts by spinning a small disc. The silk she uses for this comes from the glands that only females possess. It is neither as strong nor as elastic as other kinds of silk and as it emerges from the spinnerets, it is mixed with air which gives it a spongy texture. Once this small web is complete, she extrudes her fertilised eggs on to it and adds the sperm that she has stored in a separate sac. She carefully weaves another sheet to cover them and then joins its edges to the disc with careful over-sewing.

Araneus diadematus, an orb-web spider common in European gardens, relies on web vibrations for communication between the sexes.

Above: the somewhat smaller male ventures on to the female's web and twangs one of its threads, signalling his presence and identity.

Below: his message is received and understood and the female allows him to touch her, preparatory to mating.

In many species, the female now carries her precious parcel around with her in her jaws. Orb-web spiders fasten them to the centre of their webs. Trapdoor spiders hang them up inside their burrow. Crab spiders crouch on them and threaten anything that displays a dangerous interest in them. Wolf spiders stick them to the end of their abdomen and walk around with them.

The eggs hatch within these protective cocoons. The tiny creatures that emerge are colourless and lacking in hairs and spines. While still within the cocoon, the spiderlings moult and become miniature versions of their parents, complete in every detail except for their genitalia, but still, in most cases, smaller than a pin head.

They cut their way out of their silken home. Even then some mothers do not desert them. *Pisaura*, a wolf spider common in Europe and America, spins a larger tent for them in which they will remain until they have moulted a second time. Other mothers allow their newly emerged young to climb on to their abdomen and carry them away with them. Even at this early stage in their lives, silk is vital. They attach a silken life-line to their mother so that if they fall off they can quickly climb back on again.

Eventually they must disperse. Some simply walk away. But spiders, thanks to silk, have their own unique way of getting around. The tiny spiderlings climb to the tip of a grass stem, a twig or a leaf. They lift their little abdomens into the air and start to spin. As the silk streams from their minuscule spinnerets, even the slightest breeze catches it. The little creatures hold on tightly to their perch with their legs. Sometimes, if the breeze is sufficiently strong, they will let go and are carried up and away. If it is not, they will anchor the silk to their perch and clamber away to choose another site from which to try again.

Maybe this does not please them either and they return to the first position to see if conditions have improved. Eventually, as they cling to a filament, with yet another issuing from their spinneret and wavering in the air above them, they decide that the right moment has come. Very deliberately they turn, sever their

anchor with their jaws and away they go. Sometimes they are swept up to great heights and travel for enormous distances, across forests, mountains and even oceans.

On autumn mornings, the countryside is often draped with gossamer. Much of it will have come from spiderlings that have landed after making their first journeys. Some may have travelled no more than a yard or so. Others may have arrived after a long aerial journey of several miles. Some filaments will have been laid down by resident adult spiders as they prowled across their territories at night during their hunts. It is perhaps only on such mornings, when the low rays of the rising sun illuminate vast shining sheets of silk, that many of us can get any idea of just how successful and abundant spiders are in our countryside. A hectare of meadow may contain over a million of them.

165

4

Intimate relations

The terrestrial invertebrates have inhabited the land for well over 400 million years. It is hardly surprising that over this immense stretch of time, they have interacted with one another. But the complexity and intimacy of some of their relationships can stagger even the most inventive imagination.

Insects began to feed on plants, either dead or alive quite early in their history. The majority still do. But that is not easy. The bulk of a plant's tissues is composed of cellulose and cellulose is one of the most stable of all organic substances. It takes a great deal of processing, by both chemical and physical means, to make it digestible. Initially, the invertebrates probably relied on bacteria to do this and only consumed plant tissues when they were well rotted. Insects dealt with the problem, as did some of the back-boned animals that followed them many millions of years later, by the simple but radical step of taking bacteria into their guts where colonies of the microbes could flourish and do the work for them there.

Even so, the plants of the early forests must have been particularly hard eating. Many of the tree-ferns, horsetails and club mosses were tough and woody and grew to over 30 metres (90 feet) high. Their compressed and carbonised remains survive as seams of coal and from them fossilised fragments have been found carrying unmistakable evidence of insect damage – holes bored into trunks, leaves that have clearly been nibbled. So we can be sure that some insects had started to eat plants, even at this early date.

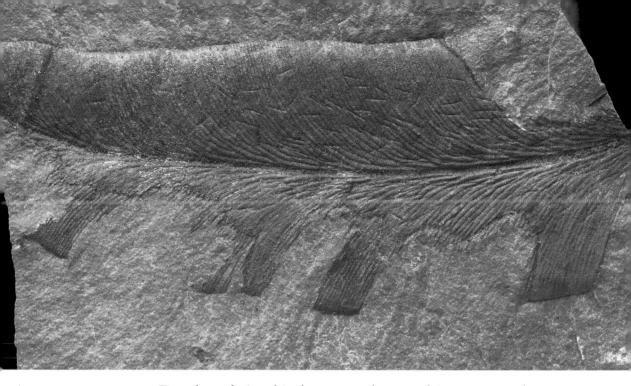

This fossilised leaf of a seed-fern from southern England provides clear evidence that insects were feeding on leaves almost three hundred million years ago.

But the relationship between plants and insects soon became more complicated. These ancient trees reproduced by first generating a small mat, not unlike one of today's liverworts, which developed on the ground around the base of the trunks and carried the male and female sexual organs. This process could only be completed in a moist environment for the microscopic male cells, which propelled themselves with small beating hairs, relied on their ability to swim to reach a female cell. This posed no problem for these trees for they grew in swamps around the fringes of the continents.

As time passed, however, and vegetation spread away from swamps and on to drier land, new kinds of trees developed that reproduced in a somewhat different way. They carried their sexual structures on their branches or on the margins of special leaves. Among them were the early conifers. They relied on the wind to carry pollen from their male flowers to the ovules in their female flowers, as their descendants do today. The method is so haphazard that if it is to be successful, a plant has to produce pollen grains in astronomical numbers. Only one grain in many millions is likely to reach a female flower. The vast bulk of it is inevitably wasted.

Among these early conifers there also grew the ancestors of today's cycads. Superficially, these plants resemble palm trees with single pillar-like trunks topped by rosettes of feathery leaves. In the centre of these terminal rosettes, they produce their cone-like structures, males on one tree, females on another. These are usually much bigger than those on a conifer. The males' are smaller than the females' but some are nonetheless a foot or more high.

◇

For a long time, it was believed that the pollen from cycads was transported by the wind, in the same way as it is from conifers. But now it has been discovered that some living cycads are pollinated by beetles. Beetles being one of the earliest groups of insects to evolve, were certainly present in those ancient forests, so it may well be that they were the first insects of all to be recruited by plants as pollinators.

The beetles that do the job today are weevils, a family which have heads elongated into a long thin snout with a tiny pair of jaws at the tip. In some species this snout may be as long as the rest of the body. With this, a weevil bores holes into wood or seeds in order to extract its food.

Every year, a male cycad produces a series of cones, each like a giant bud with its component leaves clasped tightly together. As they ripen, these special leaves, more properly called scales, swell with nutritious starch. The internal temperature of the cone begins to rise so that it is warm to the touch and it starts to give off a yeasty smell. This attracts weevils in dozens and they swarm all over the cone. Twenty-four hours later, the bud's scales begin to separate, revealing that on their underside each is carrying pollen in stupendous quantity. A single male cone may produce as many as five thousand million grains. The weevils squeeze their way in between the scales and start to scrape off the soft nutritious flesh. Then they lay their eggs on them. That done, they leave and fly away to look for another mature male cone. But they take with them a liberal dusting of the pollen that, inadvertently but

unavoidably, they collected from the underside of the cone's
scales.

The cones of a female cycad are similar in their structure and
placing to those produced by the male. In most species they are
considerably bigger. In some, they are gigantic – as much as 60
centimetres (2 feet) tall and weighing 40 kilos (90 pounds). They
do, however, produce the same chemical invitations as the males.
Consequently weevils find them equally attractive and force their
way into them. So the pollen they carry is scraped off and the
ovules, which also develop on the scales of the female cones, are
thus fertilised.

Once that has happened, the weevils' services to the cycad
have finished. The female plant offers them no reason to linger.
The scales of its cone are not as succulent as those of the male.
Worse, they contain a toxin. One quick nibble is enough to make
a weevil decide that there is no food there worth gathering and it

flies away to look for another male cone that will provide it with a better meal.

But what of the weevil eggs back in the male cone? Within a day or so of being laid, they hatch and small legless grubs emerge which eat their way into the flesh of the cone scales. If one grub meets a smaller one, it eats that as well so that after three or four days there is only a single grub to a scale. Each then constructs a shelter for itself inside which it pupates. Eventually when the cone is dry and withered, the adult weevils emerge and fly off to repeat the process. So the partnership between insects and plants that was established so long ago brought benefits to both sides.

A hundred million years or so were to pass before this relationship progressed any further. The initiative came from the plants. They developed more elaborate ways of advertising for messengers. The leaves surrounding the structures that produced their sex cells became conspicuously coloured. Flowering plants had arrived.

Water-lilies and magnolias were among the first to appear. Both today still use beetles as messengers and reward them by meals of protein-rich pollen. But other flowering plants evolved which found a way of paying their servants more economically. They debased their coinage. Instead of offering pollen, which is expensive to produce, they provided little more than sweetened water – nectar.

Individual plant families developed their pollen-saving strategies still further. Pollen delivered to a different species of plant is pollen wasted. That could be avoided if exclusive contracts were developed between particular kinds of insects and particular kinds of plant. If nectar were to be tucked away in a part of a flower that could only be reached with specialised equipment, then those insects which have such equipment will be the only ones able to collect it. They will also find it more rewarding to visit those flowers that reserved nectar for them alone. As plants evolved ways of doing this, so insects, step by step, evolved mouthparts suited to these flowers.

Beetles, the pioneer pollinators, have two sets of jaws, one in front of the other. Each is like the serrated blades of a pair of scissors. These implements chew up pollen very efficiently but they are useless when it comes to sipping nectar. Among the first insects to evolve mouthparts that could do that were the bees. Their upper pair of jaws are still scissor-like and with these they can collect pollen. But the lower pair has become modified into a structure that hinges outwards and carries a thin hair-like tongue at the end. With this, a bee can reach into many different kinds of plants to lick up nectar.

Butterflies and moths have become even more specialised. Their upper jaws have virtually disappeared so that now most are unable to eat pollen. Their lower jaws, however, have become enormously elongated and fringed along their inner margins by rows of microscopic hooks so that the pair zip together to form a hair-thin but immensely long tube. A bee's lower jaws are about 6.4 millimetres (¼ inch) long. A hawk moth's proboscis is many

The mouthparts of a beetle (below right) have acquired a serrated line of teeth on their inner edge and so become powerful cutting tools. Those of a butterfly (below left) have been turned into a tube which is so long that when not in use, it has to be kept coiled in a spiral.
▽

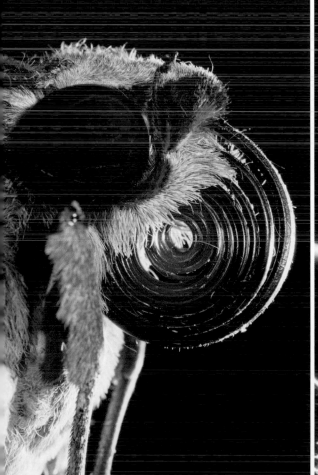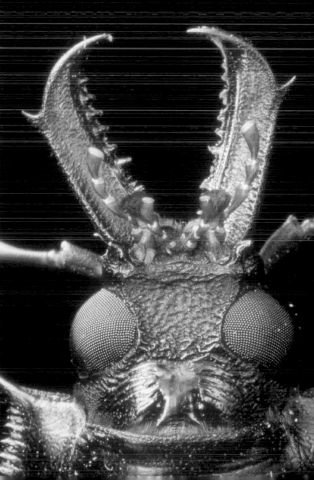

times that length. Such probosces are so long that were they to be kept permanently outstretched, they might be a considerable encumbrance in flight. That problem, however, does not arise, for the probosces of butterflies and moths contain a material called resilin which has the property, when distorted, of springing back to its original shape. The resilin proboscis of a butterfly is shaped like a watch-spring. Internal muscles can make it straighten but as soon as they relax it automatically springs back into its original coiled shape.

A bee is unable to collect nectar from a trumpet-shaped flower like a buddleia or a hibiscus in the normal way. Its fat furry body is too portly to squeeze into the flower and its proboscis is too short to reach the nectaries that lie in the flower's depths. A butterfly, however, can simply uncoil its proboscis and probe deeply into a flower with ease.

The record for the longest distance between the mouth of a flower and its nectary is 30 centimetres (nearly 12 inches). This is held by *Angraecum*, an orchid that grows in the forests of Madagascar. Its white star-shaped flowers have long dangling spurs hanging from their base and it is only in the lower tips of these that there is any nectar. The European botanists who discovered the plant were greatly puzzled as to what creature could possibly pollinate it. It was Charles Darwin, with his unique insights into the evolution of pollinating techniques, who predicted that eventually a moth would be discovered with a proboscis long enough to reach the whole distance. And so it was. It proved to be a moth called *Xanthopan* and, as might be expected, *Angraecum's* record-holding nectaries were matched by *Xanthopan's* record-holding proboscis.

It should be added that these strategies are not always and inevitably successful. Sometimes insects are so behaviourally malleable that they can break the barriers imposed by morphology. Somehow or other, bumble bees discovered that particular tubular-shaped flowers have nectar within them which is beyond the reach of their proboscis. Now they have now found ways of collecting wages that they do not earn and that were intended for

▷ *Butterflies, like this painted lady, have a tongue that is so well equipped with muscles and nerves that its owner can bend it and accurately insert it into tiny flowers to collect nectar from deep within them.*

▷ *A bee does not have such elongated mouthparts but has discovered how to collect nectar from a tubular flower by piercing its petals close to the nectary.*

others. They alight on the outside of a flower close to its base and bite a hole through it so that they can insert their tongue and, short though it is, drink from the nectaries within.

The interplay of visual signals and displays may have effects on both the perceivers and the perceived. Flowers may influence the shape of insects and insects influence the shape of flowers. Orchids, in the course of their evolution, provide spectacular examples of both reactions. The small orchids that are common in European meadows and marshlands carry spikes of diminutive flowers so elaborated with frills, lobes and hairs and so delicately patterned in browns and yellows, creams and purples that they, even to our eyes, look remarkably like bees and wasps. Wasps and bees are even more convinced than we are, for such orchids also emit a chemical attractant that precisely matches that produced by the females of a particular species of bee or wasp. When duped males settle on such a flower and attempt to copulate, they immediately receive a load of orchid pollen.

Some insects, on the other hand, imitate orchids. All mantids are hunters. They secure their prey by shooting forward the spiked forelegs that many species hold clasped in front of them, as though in prayer. For that technique to be effective, their prey has to be within their arms' reach; and to bring that about they disguise themselves. They do so to an extraordinary degree. Some brown ones hold their forelegs outstretched in front of them, with their head enfolded between them so that they resemble the bare twigs they frequent. Others are green accurately matching the grass leaves through which they clamber. Some have a remarkable ability to change within the space of a day or so from leaf green to charcoal black so that they still remain disguised even after a fire hs swept through their habitat. Most remarkable of all, a whole group of mantids are disguised as orchid flowers. One, *Hymenopus coronatus*, that habitually sits on pink orchids has pink arms and pink abdomen of an identical shade. Not only that, but the thighs of its two back pairs of legs are expanded into pink petal shapes. You may stare and stare again at a spike of orchid flowers without discerning the mantis, even when you have been told that it is there.

$\triangleright$
A beautifully camouflaged orchid mantis from southeast Asia sits in the middle of an orchid flower awaiting her unsuspecting prey.

Insects claim other things from plants as well as food. Some induce them to provide special accommodation. The body of a plant naturally contains all kinds of nooks and crannies that ants, being small and active, can easily occupy. A hollow stem, the angle of a branch, the space beneath bark, the little hollow at the base of a leaf – all these places may be colonised by communities of ants. But quite frequently plants develop special spaces that seem to have no reason for their existence except to provide a home for an ant colony.

Cordia, a shrubby tree that is found in the forests around the headwaters of the Amazon, is one of them. There is something odd about the way it grows. The South American rain forest is notorious for the extraordinarily large number of plant species it contains. They are distributed with such evenness that there may be examples of five hundred different species in as many hectares. That makes them very daunting for the novice botanist. If you identify a tree in a European woodland you may feel you are mak- ing real progress since you will soon find many more like it. But not so in the South American rain forest. There, you may flush with success at having identified one but then walk for hours, even days, before you encounter another example. Except for *Cordia*. *Cordia* trees, most unusually, often grow in what appears to be small plantations with no other species growing among them. What is more, each plantation is surrounded by a circle of earth, bare except for one or two straggly saplings that are with- ered and dying. You may well suspect that you have found an abandoned garden where someone once tried to grow vegetables. But the local people have had nothing to do with the site. Indeed, they call such patches 'devil's gardens' because they know that any plant they try to cultivate there will quickly die.

Touch a branch of a *Cordia* tree and you will immediately see the cause of these symptoms. A stream of *Azteca* ants comes run- ning out of a tiny slit in the base of the leaf stem. They are too small to bite you in any serious way, but they are quite big enough to kill any caterpillar or any other plant-eating insect that might try to make a meal from *Cordia's* leaves. They might easily make

life uncomfortable for a monkey or any other plant-eating mammal that tried to feed here.

But why are there only *Cordia* plants on the plot? And what about that ring of bare earth with its dying saplings? The ants are responsible for these too. Any seedling that manages to germinate among the *Cordia* trees and puts out a leaf is soon discovered by the ants. They swiftly swarm up it, stab its buds with the sting on the ends of their abdomen and inject them with poison. The plant wilts and dies within forty-eight hours. The same thing happens if you try to take advantage of the ring of bare earth to plant a food crop of some kind. The ants will not tolerate other species growing in their host's plot. So, steadily and inexorably, the *Cordia* trees are able to extend their territory. And in return the ants are provided with more and more accommodation.

Other ants have a different way of extending their living quarters. Colonies of *Allomerus* ants also live on *Cordia* trees, but in this case, the tree has to pay a very heavy price for its protection.

When it begins to develop its flowers, the *Allomerus* ants sting them, just as they sting any invading insects. As a consequence, the flower buds die and leaves appear instead of other flowers – and they provide extra homes for the ants. It is as if the tree is being castrated.

Another tree in the Amazonian forests, *Hirtella*, also has a close relationship with *Allomerus* ants. It provides them with living chambers in bulbous swellings at the base of each leaf where it joins the main stem. The *Allomerus* ants, true to form, castrate *Hirtella's* flowers just as they do *Cordia's*. But *Hirtella* has found a way of fighting back. As the flowering season approaches, the ants' living quarters on some branches begin to wither and the ants are forced to evacuate. The flowers on these branches can then develop unhindered and *Hirtella*, after all, is able to set seed.

△
An Allomerus *ant attacks a flower of its host, thus encouraging the tree to produce more leaves, and therefore more ant-homes.*

◇

Some two hundred different species of ants are known to take part in partnerships like these. And plants provide them with lodgings within their roots and stems, in their trunks and branches, beneath little tents on the surface of leaves and between curling flanges on the sides of stalks.

One of the most impressive collaborations can be found in the mangrove swamps of southeast Asia and New Guinea. Here and there, hanging above the arching mangrove roots, clasping a spindly trunk or branch with a network of roots and looking surprisingly green and healthy in this otherwise bleak tangle, you will find *Myrmecodia*, a spiny globe, the size of a small football, with a bouquet of leaves sprouting from one side. And running all over its surface, in between the spines and in and out of holes, are small black ants.

Cut open the globe and you will see inside a maze of interconnecting chambers and passageways. Some of these have a smooth, white interior. Others are lined with a brown sediment and covered with small rough warts. These are the ants' latrines and waste dumps. The ants not only excrete here, they deposit the remains of the insects on which they feed. The brown lining is consequently rich in nitrogen and phosphorus which the plant is able to absorb and which are exactly the elements that it needs for healthy growth. The ants, in short, are feeding the plant. And in return they are provided with the most spacious of all vegetable ant-mansions.

In some instances, the nutritional balance is tilted the other way. The plants provide special food for the ants. Acacias in both Africa and South America are armed with extremely large needle-sharp spines that make brushing against them a very painful business. The bases of these spines are bloated and hollow and inhabited by very aggressive ants. Grazing animals, shaking the branches as they try to nibble at the leaves, not only risk scratching their delicate muzzles but also get badly stung by the ants if they do not move away within a very short time. The acacias reward their defenders not only with lodgings but with special food. Small orange-coloured fat-rich granules sprout from the tips of

their branches. These are neither flower buds nor fruits but payments for their army of mercenaries. The ants collect them, chew them into small pieces and then feed them to their developing larvae.

The question must be asked as to whether the acacias really do benefit from the presence of ants on their branches, or whether they are simply exploited by the ants and get nothing in return. Some evidence comes from other species in their family. Acacias are found, not only in Africa and South America but in Australia. That continent, of course, has virtually no large animals that live by browsing. Kangaroos and others of their kind are primarily grass feeders. And Australian acacias, significantly, have neither thorns, nor ants living within them.

There is also experimental evidence. In South America, individual trees were stripped of their ants and kept free of them for many months. Bugs arrived and sucked their sap, beetles gnawed

Acacia ants gather the fatty nodules specially grown by a South American acacia to reward them for their services as guardians.

into their branches, caterpillars chewed up their leaves. Seedlings sprouted nearby and began to overshadow them. After a year the ant-less acacias had become severely stunted and were not sufficiently vigorous to produce any seeds whatever.

There is one further question. Would the acacias produce homes and food for ants whether the ants were there or not? That is difficult to decide from observations in the wild for ants are so pervasive in most environments that all plants offering such accommodation are always occupied. But again, experimentation can produce evidence. Acacias, even when grown in ant-proof locations still develop hollow swollen thorns. That suggests that the plants' relationships with ants are so long-standing that instructions to produce the facilities ants need have now become fixed in the acacia's genetic make-up.

An ant mansion, produced by an African acacia to accommodate the many ants that swarm over its branches ready to attack the muzzles of hungry herbivores.
▽

Some insects, however, have a mysterious ability to compel a plant to produce structures that are greatly to the insects' benefit but bring no advantage whatsoever to the plant itself. It is as if such insects are able to genetically engineer a plant's tissues and instruct them to grow in quite different ways. Structures produced in this manner are known as galls. Many different groups of animals possess this strange and still little-understood ability. Nematode worms, mites and bacteria are among them. But insects exploit the technique more extensively than any other group. Thirteen thousand different species of them do so. They include butterflies, beetles, wasps and bugs. Two large and entire families – gall wasps and gall midges – can live in no other way. The galls they induce vary from leaves contorted into folds or blisters to large conspicuous structures that look like some kind of strange fruit. One of the most familiar is the bundle of tangled red hairs growing in European hedges on the stems of wild roses that are known as Robin's pincushions.

△
Robin's pincushion, a common gall that develops on a wild rose plant when one of its buds is injected by a tiny gall wasp.

▷
Two of the many galls that may afflict an oak tree – above, a spangle gall and below, a silk button gall – both the result of gall wasp activity.

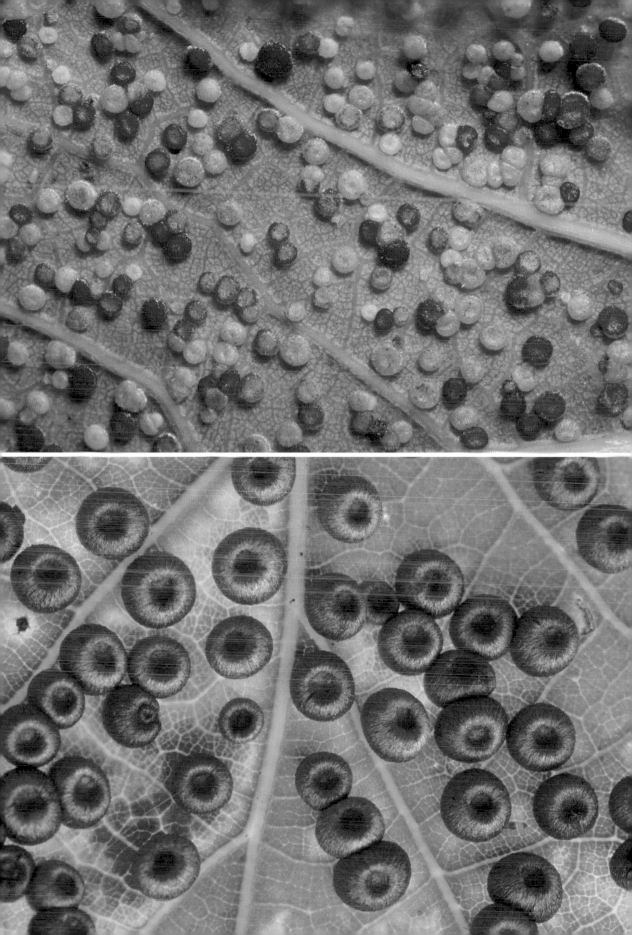

Oak trees are particularly susceptible to this kind of genetic subversion. The oak genus as a whole contains about fifty different species and between them they carry over a thousand different kinds of gall. Any one oak tree may be afflicted by as many as seventy different kinds, spangling the underside of its leaves, bulging like warts from their surface, and encrusting its acorns.

These galls are caused by the activities of small black wasps, less than a centimetre long. Each species creates its own particular kind of gall. In spring, a female alights on a part of the oak where its tissues are about to begin rapid development – on the bud of one of its flowers or leaves, the tip of a shoot or the underside of a young leaf – and with the hair-thin ovipositor at the end of her abdomen, injects a microscopic egg. As soon as this hatches, the character of the plant cells around it begins to change. The tiny grub that has emerged from the egg has started to secrete a mysterious substance that has still not been identified chemically, which causes the cells to develop not into leaves, flowers or anything else that the oak normally produces. Instead it grows into a gall.

The wasp grub lies within it as the gall grows, feeding on the special cells that line its interior. These are very nutritious, easily digested and lack the poisonous toxins with which the oak protects many of its other tissues. As fast as the grub eats this excellent food, the oak renews it. So unlike other insects which feed on the oak's leaves and have to be continually on the move to find them, the wasp grub has induced the oak to bring food to it, packaged in easy-to-eat mouthfuls.

Marble galls are small and spherical, about two or three centimetres (an inch or so) in diameter, green when young and eventually brown. Their shells are hard and provide the wasp larva growing inside with very good protection. But they are not impregnable. Woodpeckers can split them with a blow of their beak in order to eat the grub. Squirrels gnaw them open. But the greatest danger to a grub lying in its globular fortress comes from another female insect belonging to a different species of wasp.

▷
A gall wasp, less than a centimetre long, about to inject the bud of a dog rose in spring. England.

This one is rather larger than the female that injected her egg into the oak bud in the first place. She has an even longer ovipositor and one that has a mobile tip equipped with sense organs. She drills into the gall with this extraordinary piece of equipment and with it probes around the interior searching for the grub, guided by its faint smell. When eventually she locates it, she delicately ejects an egg which sticks to the skin of the grub. So soon there are two grubs within the gall. But the newcomer does not eat the oak tree's tissues – at least not directly. Instead it eats the original tenant, remorselessly nibbling at its vitals. Eventually, with most of its tissues gone, the original grub dies. The bloated interloper now pupates. When at last the adult wasp emerges from its pupal skin, whether it is the first occupant or the second, it bores its way out of the gall and flies off to mate and then repeat the whole process.

Some galls however, protect themselves from this kind of intrusion. One kind forms clusters of small drop-shaped growths

△
A gall wasp injects her egg into the existing gall so that her larva may feed not only on the gall's tissues but on the tiny larva of a different species that is already within the gall.

that exude a sweet nectar from their surface. This attracts ants. Groups of them are always around on oak trees. They soon discover the sugary fluid and stay on the gall lapping it up just as fast as the gall produces it. If a parasitic wasp alights on such a gall in order to lay her egg, then the ants attack her until she flies off.

This is particularly remarkable for it is not, of course, to the advantage of the oak tree to protect its galls in this way. On the contrary, it might be better for the tree to get rid of them if it could do so. Nor does an oak produce nectar under any other circumstances. It has no need to attract insects to carry away the pollen from its catkins for it relies on the wind to do so. The organism that benefits from the production of nectar is the original creator of the gall, the wasp. Having induced the oak to provide it with food and shelter, it has now stimulated the oak's tissues to make the tree produce a substance that it never normally manufactures to provide wages for the wasps' guardians. Gall wasps developed ways of genetically engineering plants long before humanity did so.

A gall wasp grub in its feeding chamber in the centre of a cherry gall on an oak leaf.
▽

The animals of the undergrowth have not only struck up partnerships with plants but with one another. Just as acacia ants have somehow induced trees to provide them with a specially developed crop, so other ants persuade some kinds of insects to do the same. And they care for and exploit them in much the same way as human beings treat their domesticated cows.

The cows in question are aphids and the milk they produce is known as honeydew. Aphids – and there are some two thousand species of them – are sap drinkers. They include the greenfly which so troubles gardeners. Some live on the stems of plants, others on their roots below ground. They insert their long thin mouthparts into the sap-carrying vessels of a plant and drink. The pressure of the sap within the plant is sufficient to drive liquid up the aphid's mouthparts and into its stomach. The aphid has to make little if any effort to get its food, apart from plugging in its stylus in the first place.

Sap is very rich in sugars and very poor in nitrogen. Nitrogen, however, is an essential element in an insect's diet and to get enough the aphids have to drink a great deal of sap. They excrete the excess by squirting it from their rears or allowing it to accumulate as a slowly growing drop. It still contains a great deal of sugar. The aphids may not need that, but ants do.

Just like human farmers ensuring that their dairy herds get the best pasture, some ants drive the aphids to places on a plant that are particularly suitable for sap collection. In bad weather, they will build small byres on plant stems, using particles of soil and leaves, in which the aphids can shelter. As droplets of honeydew accumulate on the tips of the aphids' abdomens, so the ants collect it. And very productive their charges are. Some species of aphid can excrete well over their own weight in liquid every hour. What is more, some aphid species produce as much as three times more honeydew if regularly milked by the ants than if they are left alone.

The ants also help to propagate their herds. When the aphids produce young, which are born alive, the ants escort or even carry the youngsters to suitable feeding places on the plant. Some ants have been observed to mark their charges with substances that are

An ant tends its flock of sap-sucking, honeydew-secreting aphids.

specific to their own particular colony, the equivalent perhaps of branding cattle to warn off poachers. And they guard them against predators. Ladybird larvae are voracious little carnivores and will attack caterpillars and many other kinds of plant-eating insects. But they stand little chance of success if a flock of aphids is protected by ant shepherds.

And finally, just as *Allomerus* castrates the *Cordia* trees, so there is some evidence that aphid-tending ants interfere with the breeding of their charges to increase the size and productivity of their herds. When an aphid population rises above a certain level, it will fragment. As individuals moult, some emerge as sexually mature winged forms and fly away to mate and establish new groups. That, of course, would not suit the ants and to some extent, they prevent it happening. It seems that they secrete a hormonal substance of some kind from glands within their mouths. This is transferred to the aphids as the ants drink their honeydew and prevents the onset of the aphids' sexual maturity. So the ant farmers increase both the productivity and the size of their herds to levels they would not otherwise reach.

◇

Aphids are not the only sap-feeders to be milked by ants. Some treehoppers, which also drink sap, are treated in just the same way. In southern Brazil, a particularly colourful treehopper species feeds on fruiting mango trees and is such a valuable and productive milch-cow that ants give small herds of them round-the-clock protection, one ant species guarding them all day, and another nocturnal species taking over the responsibility – and collecting the benefits – throughout the night. They even guard treehopper eggs if the female that lays them happens to stray away from them.

Ants are not alone in having discovered this valuable supply of food. Small bees, of a kind that normally feed largely on resin, also tend the hoppers. On some mango trees, it is not unusual to see some stems being patrolled by bees while others are guarded by ants. Of the two, the bees are the more agile. If an ant strays from its stem on to that guarded by resin bees, the bees will buzz

▷
The flag-legged bug waves the flanges on its hind legs to warn off predators, but other closely-related species display them in order to attract their prey.

around it aggressively. If it persists, they will even physically grab the intruder and carry it off to dump it on the ground far away from the tree.

Bugs are related to treehoppers and have similar mouthparts modified into piercing stylets. One of them, the feather-legged bug, also has a relationship with ants – but not always to the ants' benefit. This species belongs to a closely related group which have extraordinary flanges on their hind legs. Some of these, including the flag-legged bug, brandish them to warn off predators. But the feather-legged uses them to signal to ants. It waves them energetically in the air. Small ants of many kinds are attracted by these signals and approach the bug to investigate. The bug encourages them. It lifts its thorax and reveals a gland on its chest. The ants start to lick it, but within a few seconds, they fall over as if in a trance. The bug's secretions contain some kind of anaesthetic. The bug then turns to them and plunges its stylet into their vitals and sucks them dry.

◇

Insects have an extraordinary facility in imitating the appearance of one another. In some instances, this mimicry can be so perfect that even the most expert entomologist may be misled. There may well be instances where the deception has still not been detected and two quite separate species are still regarded as one by science, with DNA analysis the only tool likely to reveal the confusion. Harmless species frequently mimic a poisonous or ferocious one. Bees and wasps have stings. What is more, they advertise the fact by wearing warning colours – black and yellow stripes. A single attempt to eat one is enough to persuade an innocent bird that it is unwise to do so. Most insect-eating birds, therefore, leave such creatures alone. But harmless creatures, if they assume a similar vivid uniform stand a very good chance of also being ignored by a bird, no matter how hungry it is. The system works as long as the truly dangerous model remains more numerous than the harmless mimic.

So it is that the hornet clear-wing moth (*Sesia apiformis*) has indeed come to look like a hornet, and yellow-striped wasp-beetles

▷
Model and mimic. Above: a true hornet drinking. With its powerful sting it is avoided by most insect-eating birds.

Below: a harmless moth, the hornet clear-wing, which most birds also avoid. Freshly emerged, it has not flown yet: when it does, it soon loses its wing scales, to become a 'clear-wing'. England.

(*Metoecus*) to look like very wasps. Tiger beetles, which have a poisonous liquid in their body are mimicked by other beetles that are perfectly digestible, and cockroaches develop red wing-covers with black spots so that they resemble those ferocious and bitterly distasteful beetles, ladybirds.

One of the most extreme of all disguises is that developed by *Heteronotus*, a small sap-drinking treehopper from the tropical forests of Central America. Viewing it from above, as a bird might do, it looks remarkably like an ant, albeit one that seems to favour walking backwards. It appears to have a formidable pair of gaping white-tipped jaws, a narrow waist and a glossy black abdomen. But this is the model of an ant, a bulging barbed extension of the bug's thorax. Underneath it, facing in the opposite direction and largely concealed from above the bug has an abdomen and membranous wings which are wholly typical of its family.

▷
Some treehopper species adopt elaborate disguises. Above, the bulbous outgrowth contains no vital organs but screens the hopper's abdomen beneath. Below, this one has spines and spikes, like a black ant with its head facing the hopper's tail.

A south-east Asian spider has a green abdomen, helping it to get past fierce green tree ants and prey on their larvae.
▽

Small spiders, which have soft succulent bodies, are preyed upon by many birds. They too could benefit from disguising themselves as poisonous insects. But they have a particular problem. Spiders, after all, have only two sections to their bodies and eight legs, whereas insects have three body sections but only six legs. A jumping spider from Borneo, *Myrmarachne*, has dealt with the problem by developing a groove – a false waist – across the middle of its cephalothorax. And it conceals its possession of an extra pair of legs by holding its first pair, which are longer than the others, in front of its head and waving them around in much the same way as an ant waves its antennae. The effect is so convincing that even an experienced eye may be deceived – until, that is, the supposed ant, in trying to escape, drops from its leaf and descends to the ground on a filament of silk.

Above: the back-to-front jumping spider mimics ants with its colour, a bogus waist, and two rear spinnerets resembling an ant's antennae. Borneo.

Below: the only European ant-mimicking spider. France.

Sometimes, however, both partners in a mimicking duet are poisonous. Both have adopted warning colours in patterns that are so similar that only the most detailed examination can tell them apart. This is particularly common where the partners are, in any case, closely related. Such is the case between butterflies belonging to the genus *Heliconia*. In some instances however, the two butterfly species concerned come from very different families, yet the patterns on their wings are almost identical. In mimicry of this kind, it is not just one partner, the harmless one, that benefits. Both do. The more widespread the warning, the more universally recognised it will be. If another species helps in propagating the meaning of the message, then so much the better. The system even benefits the predator, for a single distasteful meal is enough to convince it that individuals of not one but at least two species are not worth eating.

◇

In due course, over the long passage of geological time, amphibians and reptiles, birds and mammals appeared on land. Many preyed upon the resident invertebrates, but the invertebrates themselves were not entirely defenceless or passive. Many found ways of turning the tables and feeding on the giant creatures that had come to live among them. Relatives of the spiders did so,

197

probably at a very early date indeed, for today they parasitise every kind of land-living vertebrate as well as others among their fellow invertebrates. These are the mites.

Most, when adult, still retain the eight legs of their arachnid ancestors. Nearly all are tiny. Some are detritus feeders, some live on vegetation and some are hunters. A cubic metre of fertile soil may contain as many as a million of them belonging to several hundred different species. Those that are parasitic are not necessarily harmful. Some in fact are positively beneficial, for they rid their hosts of excess fatty secretion, waste skin and other detritus. One kind is often seen on bumblebees. A queen in spring may carry up to a hundred of them, clinging to her fur. They are not feeding on her body, merely hitching a lift, for their ultimate home is within a bumblebee's nest where they are useful scavengers. She is obligingly transporting them from their old home to a new one.

The biggest — and therefore the most familiar to us — of all mites are the ticks. Some have soft leathery bodies, live in their hosts' nests and burrows and only attach themselves to their land-lords when they need to feed on their blood. Others however, sink their jaws in their hosts and then remain attached. These have bodies that are protected by a small shield. Initially, this almost covers their abdomen, and gives them such efficient protection that they are known as 'hard ticks'. But as they feed so they grow until eventually they may measure 2-3 centimetres across (almost an inch) and their protective plate becomes nothing more than a small scale attached to the back of their heads in front of a grossly swollen abdomen. So a hard tick may gorge itself for weeks on blood, wedged between a reptile's scales or clinging to the skin between a bird's feathers, hanging conspicuously from a mam-mal's eye-lids or out of sight and reach, up a nostril or down an ear. Eventually, fully sated, it drops off either to moult or to mate, or if it is an adult female, to shed its eggs.

◇

Above: a hungry tick, searching for a meal.

Below: clinging between the hairs of a mammal's coat, this tick is now full of its host's blood.

Insects too managed to extract meals from mammals. Fleas, rela-tives of the flies, began to explore the hairy coats of mammals and there feast on blood. They drink it through a microscopically thin stylet that they drive into their host's body with a muscular mech-anism inside the head similar in principle to a trip-hammer. There are no fleas on amphibians or reptiles, so it seems that the ancestral fleas can only have embarked on this way of life when mammals first appeared some 65 million years ago. Today, there are about 2,500 different species of them. Of these, about a hundred live on birds, which suggests that their move from mammals to this new kind of host happened comparatively recently.

Adult fleas have become extremely adapted to their life among feathers and hairs. Wings would be an encumbrance in such cir-cumstances and fleas have lost them though some species still retain the remnants of wing buds on their thoraxes. Instead of fly-ing, fleas jump. Their hind legs have become greatly enlarged and equipped with powerful muscles that enable a flea to hop

201

prodigious distances. A cat flea, for example can leap about 30 centimetres (about 12 inches) into the air.

▷

A cat flea jumping. Having lost their wings, fleas rely on their long jumping hind legs when moving from one host to another.

A flea's head resembles the bows of a ship with a blade-like leading edge with grooves in it into which the antennae neatly slot. Its body is flattened from side to side, so that it can slip easily between its host's hairs and is equipped with lines of widely-spaced spikes and bristles. These all point backwards so that if the insect moves in that direction, either of its own volition or because it is being pulled by an irritated host, they engage with their host's hairs and the flea becomes very difficult to dislodge. And the carapace itself is so strong, thick and slippery that even if a host does manage to get a grip on a flea, it is not easy to crack and kill.

These adaptations are so extreme that they tend to cloak any difference between different species. Female fleas, in fact, are extremely difficult to tell apart. It is primarily the details of the males' complex genitalia that enable flea scientists to classify them.

A few fleas live permanently on their hosts. One or two species fasten themselves to hairs or feathers. The majority, however, spend much of their time in their hosts' nest or bedding and only hop on to the owner when they are hungry. So it is that fleas tend to afflict only those mammals that have permanent burrows and dens – rats and rabbits, cats and dogs, even bats. Apes, which in the wild do not make nests or only occupy them for a night or so, do not have fleas. It is likely that human beings acquired the species of flea that is now regarded as their special possession from badgers or pigs. But they probably only did so when they abandoned the nomadic life and settled down in homes of one kind or another.

◇

Attending to the safety of eggs and providing the young larvae with food, which may be quite different from that taken by the adults, can be a major chore for any animal. Many insect parents, of course, ignore it. They simply abandon their larvae to look after themselves and compensate for the massive and inevitable mortality that results by laying vast numbers of eggs in the first place. But

many other species have evolved ingenious ways of imposing these responsibilities on others. The range of techniques they have developed is extraordinary.

Stick insects have bodies, legs and antennae that are so slender and elongated that it is extremely difficult to detect them among tree twigs – unless they are moving. The females have reduced the whole business of reproduction to its basic and simple essentials. Most of them, for most of the time, dispense with the help of the male. The eggs they produce are already fertile and simply drop to the ground as the females clamber about in trees, feeding.

One Australian species eats almost nothing but the leaves of casuarina trees. The eggs produced by the females are small, round and sculpted with delicate ridges. To the human eye, they look remarkably like the seeds of the casuarinas. And not only to the human eye. Casuarina seeds have small fleshy oil–rich attachments which are very nutritious. Accordingly, harvester ants collect them with great assiduity and take them back to store in their nests and later consume. But they, just like humans, find it hard to distinguish between casuarina seeds and stick insect eggs. So they collect both with equal enthusiasm and deposit them in their underground granaries. It is only when they come to feed on their stores that they notice that all their collected grains are not the same. The insect eggs don't have the nutritious attachments carried by the seeds. So they leave them uneaten and the eggs are able to mature and develop in safety underground.

The mimicry may not end there. The infant stick insects when they hatch look very like newly emerged ants. Whether this resemblance is fortuitous or not, the ants allow them to walk out of their nest and up on to the casuarinas to take their first meals of leaves.

◇

Bot–flies parasitise cattle and other ungulates. In many species, the female fly lays her eggs on the hair of her victim. In one, the human bot–fly, the female astonishingly uses insect messengers to place them there. She pounces on another species of fly, such as a

A bot-fly must be able to fly with the greatest accuracy if it is to fix its egg on another insect in mid-air and accordingly has extremely large eyes which give it the necessary visual acuity.

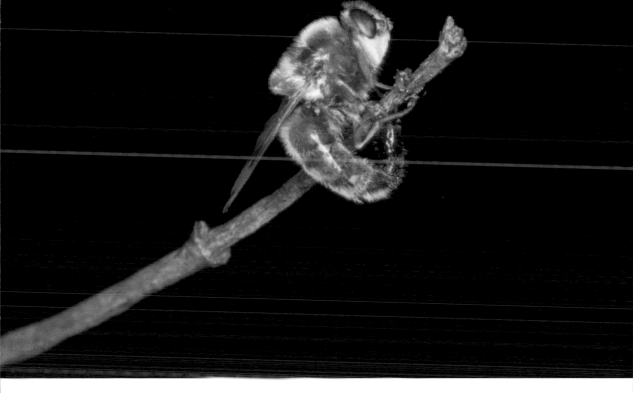

deer fly, a stable fly or even a mosquito. Occasionally the two, coupled together, will fall to the ground and there the bot-fly glues a single egg on her press-ganged messenger. Sometimes she achieves this feat while the two are still in mid-air. Stable flies may carry as many as a dozen bot-fly eggs cemented to them but surprisingly, bearing in mind the difficulty of placing them there, each will have come from a different bot-fly female.

Why don't the female bot-flies lay their eggs directly on the animals whose flesh they will consume? Perhaps that is because their hosts might well notice such a big fly landing upon them and get rid of them before the eggs are laid.

Once the eggs arrive in a mammal's coat they may be licked off by their host and so get into the victim's throat. More frequently the little larvae, as soon as they hatch, bore their own way through the animal's skin. The human bot-fly larva creates a small feeding pocket for itself just under the skin and stays there – unless its host manages to extract it. The larvae of other species wander around between the various internal organs and muscles of their hosts, feeding as they go. Eventually, they find their way back to the animal's outer regions and form little capsules beneath its skin. After five days, having grown fat and large, their development stimulated by the warmth of their host's body, they bore their way through the uppermost layers of the skin and fall to the ground. There they pupate.

◇

Wasps are among the most ferocious and lethal carnivores in the invertebrate world and many of their species also provide living flesh for their young. A large *Plesiometa* spider, hanging from the centre of its huge web is something most insects avoid. But not a female *Hymenoepimecis* wasp. She hovers in front of the spider for a few seconds and then lands on her. Quickly, she brings forward her ovipositor and then, almost immediately, flies off again. The spider seems no worse from the encounter. But a wasp egg has been implanted on her back. When it develops, the young maggot remains there, riding pick-a-back, while it continues to absorb nourishment from its carrier's body.

The night before the larva pupates, the spider starts to destroy her web, as most web-spinning spiders do, eating the silk so that she may reprocess it. But then, at midnight she starts to spin a new one. What she produces now is not her normal orb-web. It has no radial spokes or sticky spirals and it is secured to the surrounding vegetation by particularly strong threads. It will be both her death shroud and a cradle. Once it is finished, she crouches motionless beneath it. She will not move again. The wasp larva has injected her with a substance that first changed her web-spinning behaviour and now kills her. The larva sucks out every drop of nutriment from her body, using her legs like drinking straws. Then around dawn, it drops the spider's husk and starts to spin its own bright orange cocoon which will hang, safe from ants and the rain, inside the construction that the spider created with her last dying efforts.

◇

▷
A giant Plesiometa *spider, powerful and ferocious though it is, cannot detach the larva of a* Hymenoepimecis *wasp which is clinging to its abdomen and eating it alive.*

Bee-flies are so-called because they do indeed look very like bees. Their true character only becomes obvious when you see them hover in front of a flower or dart away at high speed with an agility that is wholly fly-like and not in the least like the rather lumbering flight of a bee. But they also merit the name for a less obvious reason. The female of one of the European species lays her eggs directly on the ground but close to the nest of a solitary bee. There they hatch into slim reddish-yellow worms which wriggle their way with great vigour into the bee's nest. Once there, they moult in the same way that most grubs do but, more unusually, they emerge as rather different-looking grubs – short squat maggots that are much less active. These proceed to eat the stores of pollen and honey that the bee has so laboriously collected for its own young. When these foods are finished, the maggots set about the bee's own larvae and consume them with equal gusto.

In the deserts of the American southwest, this contest between bee-fly and bees has been taken a stage further. Species of bee here feed primarily on the pollen which they collect from the flowers of prickly pear and the huge pillar-like saguaro cactus. They nest singly in holes which they excavate with great energy in the sandy ground. Where a site is particular suitable, there may be tens of thousands of them nesting in a patch of desert the size of a tennis court. Having dug a tunnel, which may be up to 20cm (8 inches) long, they construct cells in each of which they lay an egg and stores of pollen and honey. But they also fortify the tunnel against bee-flies. Some species use mud mixed with a little regurgitated honey to construct a small turret over the entrance. Others build the entrance with a horizontal tube along the ground which extends for several inches.

But the bee-flies have found a way of storming these defences. They perch on grass stems or pebbles near the bees' nesting grounds. As a bee leaves its nest to collect more pollen, the bee-fly takes off and hovers just above the entrance of a turret or a tunnel. Then, rapidly flicking its abdomen forward with a series of jerks and while it is still in mid-air, it fires eggs into the entrance. Many of its eggs don't reach the burrow itself. But some do. Once

▷
Bee-flies collect nectar from flowers. In the lower picture the bee-fly is perched in profile, so the large brown object above its outstretched tongue is not its head but one of its huge eyes.

inside, after a short pause, the fly's eggs hatch and the larvae crawl down into the bee's nesting chamber. Unlike the European species, they ignore the food stores. Instead, each immediately attaches itself to the flanks of a newly hatched bee larva. As this feeds on the provisions left by its mother, so the fly larva sucks out its vitals until all that is left is a shrivelled husk.

The proportion of fly eggs that manage to reach a bee's nest chamber is not high. But the strategy is so successful that there may be as many as five different species of bee-fly working the nesting ground of a single bee species.

◇

Blister beetles have devised a rather more complex way of conscripting foster parents for their young. Indeed, it is so complex and incurs such heavy losses that one wonders whether it really is a better strategy than simply leaving the grubs to their own devices. Only a tiny proportion of eggs succeed in completing a blister beetle obstacle course and the female, in consequence lays as many as ten thousand of them. But only two need to survive to maintain the size of the blister beetle population.

One species of these beetles *Meloe franciscanus* lives in the Mojave Desert in the western United States. Females lay their multitude of eggs on the surface of the sand. These hatch into tiny bristly black creatures that are so unlike their parents or indeed any other beetle larvae that for a long time they were thought not to be beetles at all but lice. They stay close to one another and, moving as a group, they ascend the nearest grass stem and there cling together in a single shining mass. These assemblages attract the males of a kind of desert bee, *Habropoda pallida*. There may be some visual resemblance to a female bee. Certainly, the size of the group as a whole is about the same, but to the human eye there is little other similarity. It is more likely that the group emits one of those pheromones, a perfumed invitation, that matches that released by a female bee ready to mate.

However that may be, a male bee is likely to appear within only a few minutes of the tiny larvae taking up their position. He

▷
Above: a cluster of newly hatched blister beetle larvae, waiting on a grass stem for the visit of a male bee.

Below: a male desert bee, with a load of beetle larvae on his back feeds on nectar, fuelling his search for a female and the shedding of his load.

hovers within a few millimetres of the cluster and prepares to land on it in exactly the same manner as he would land on a female bee. Immediately – and so swiftly that the action can only be followed in any detail with a slow-motion camera – the entire aggregation transfers itself on to the male bee's body. He, since his putative mate has inexplicably vanished, then flies off wearing what appears to be a small black jacket.

So far, so good. But this is only the first hazard in the beetle larvae's obstacle course. The male bee will only carry them for one further stage along it. Before long, with luck, he detects once again an alluring pheromone. A genuine female awaits him, clinging to a grass stem. He alights on her back and prepares to copulate. Immediately, all the little beetle larvae, moving once again with remarkable synchronicity, transfer themselves on to the female.

They are now on their last lap. The female, having been fertilised, returns to her nest in a hole in the ground, carrying them with her. Inside, there are open cells some of which she has filled with pollen. Swiftly the beetle larvae drop off her, and there, in the shelter of her nest with ample stores of food at their disposal, they feed, grow and eventually pupate.

◇

Sometimes several of these relationships become interlocked. Complexity is piled on complexity.

Blue butterflies are not very big but seeing them in a wild meadow fluttering from flower to flower is one of the great rewards of a paradisiacal European summer. However, blues have a devious side to their characters – at least during their youth – that belies the gentle romantic character we are inclined to attribute to them.

The Alcon blue butterfly of central Europe lays her eggs on the leaves of a gentian plant. One might expect that like most butterfly caterpillars, the little larvae when they hatch will start to nibble the gentian leaves. But not so. They are small oval creatures, more like miniature wood-lice than the spectacular multi-legged

▷
An ant working in its nest's nursery, tends the larvae. Among them, however, lies the slightly larger, pinker larva of a blue butterfly.

sausage-shaped leaf-munchers that the word caterpillar normally brings to mind. They burrow into the gentian buds and feed. After some time, having grown considerably, they drop to the ground and there they are discovered by ants. At this point the caterpillars of other species of blue butterfly excrete a kind of honeydew from their flanks which the ants clearly relish for they pick up the caterpillars and take them back to their nest. The Alcon blue caterpillar however, does not bother to do this. Instead it simply emits a chemical signal which instructs the worker ants to treat it as though it were one of their own larvae. Accordingly, they pick it up and carry it back to the complex of tiny galleries and chambers that they have excavated around the roots of a clump of grass. There, they feed it.

So potent are the caterpillar's chemical instructions that the ant workers do even more. They give it preferential treatment. If you pull up the grass tussock to look inside the ant nest you can watch the ants, in the face of this disaster, disregard their own larvae and pick up the butterfly caterpillar to rush it away to safety. The ants continue slavishly to feed the young butterfly for the next two years until eventually it is fully grown and ready to change into its adult form.

At this point, a new character in the drama may appear – a female wasp. She lands on the ants' grass tussock and clambers down towards their nest, feeling around with her antennae. In some way that no one understands, she is able to detect whether or not a particular nest contains a butterfly larva. If it does, she then clambers farther inside. The ants panic, blindly rushing around the intruder trying to bite her and carry her away. But now the wasp releases her own chemical message which not only repels the ants but causes them to attack one another. In the midst

One of the loveliest of central Europe's butterflies, the Alcon blue that plays such a sinister part in the lives of ants.
▽

of this confusion, the wasp walks even deeper into the nest and eventually locates the caterpillar. And then she injects it with an egg.

The butterfly larva remains as active as before and appears to the devotedly servile worker ants to be just as hungry. So they continue to feed it as assiduously as ever. The food they bring, however, goes not to sustain the ants' original foster child but the wasp grub that is now growing inside it. In due course, the caterpillar starts to pupate and soon disappears from sight within its brown pupal case. But what will emerge the following spring is not another Alcon blue butterfly. It will be a wasp.

In the whole of the animal kingdom there are no life cycles more complex and improbable than those that involve insects. One might well wonder why. There may be two explanations. First, the life cycle of most insects is very brief. Fruit flies can produce a new generation every ten days. Each occasion they do so presents an opportunity for some variation to appear in their genetic make-up which can be the seed of a new adaptation. So evolution is able to proceed many times more swiftly among insects than it can among most amphibians, reptiles, birds or mammals, most of which only produce a single generation in a year.

But secondly, insects appeared so early in the history of life on land that they started on their evolutionary adaptations many millions of years before any backboned creature did so. In short, the invertebrate animals of the undergrowth have had more time to evolve and are able to do so at much greater speed, than any other group of land-living animals.

5

Supersocieties

The insects' external skeleton of chitin has served them very
well. Beetles have strengthened it to produce entomological
equivalents of armoured tanks. Butterflies have thinned and
stretched it into wings that are strong and rigid yet feather–light. It
can be made impervious to liquid and thus help an insect to retain
all its body fluids. It can be coloured for camouflage or embla-
zoned with the most vivid metallic colours in the whole of the
natural world. It can be jointed to give it flexibility and by developing internal spikes and bulges that serve as muscle attachments,
it can be motorised so that an animal can skip or run, hop or leap,
burrow or fly. If damaged it can be given a temporary internal re-
pair and then, at the next moult, be re-grown not only slightly
larger but once again complete and perfect.

But that external skeleton may also be the reason why no insect
grows to the size of an antelope or a wolf, let alone a dinosaur or
an elephant. Since it cannot be expanded, it must be regularly
moulted if its owner is to grow. In between shedding the old skeleton and hardening a new one an animal is, necessarily, without
any rigid support for its body. That could not only make it fatally
vulnerable and defenceless, but incapable of maintaining its shape.
An animal the size of an antelope, if deprived of any kind of skele-
tal support, would simply slump and flop.

An alternative explanation for the apparent restriction on insect
size is connected with the way they breathe through the thin
branching tubes, the tracheae, that penetrate every part of the
body and carry oxygen directly to the tissues. Oxygen can only

diffuse along the tracheae at a comparatively slow speed. This may not, however, be the crucial limitation. Insect breathing need not be entirely passive. Bees are able to inflate and deflate their abdomens and so hasten the movement of air up and down the tracheae. And it is hard to believe that the evolutionary processes that can produce such complex structures as multi-faceted eyes and hypodermic stings could not also have produced more effective ways of breathing had this been the insects' major barrier to growth.

Whatever the cause may be, however, insects have found a way of transcending any disadvantages there might be in being small. In some species, great numbers band together and so coordinate their activities and divide their tasks among themselves that they behave as if they were one giant super-organism. In that form they can dominate their environment just as effectively as a plant-eater like an antelope or a hunter like a wolf.

The first indications of how they might manage this may have become visible as long as 300 million years ago. At that time, cockroaches were one of the great insect successes. They still thrive today. They have acquired a bad name because several species have found ways of sharing our houses and eating our food. There are, however, some four thousand others that live in almost every habitable environment, in forests and tundra, mountains and swamps. They are flattened creatures two or three centimetres (an inch or so) long, with no waist between their thorax and abdomens so that their outline is a smooth uninterrupted oval. They have leathery fore-wings that fold back and protect a more membranous and similar-sized rear pair. And with efficient, all-purpose mouthparts, they can eat almost anything.

Many species feed on that abundant and intractable material cellulose, which they collect in the form of leaves, grass or dead wood. They digest it with the help of protozoan micro-organisms, called flagellates, colonies of which flourish in their guts. When their eggs hatch, each larva must somehow acquire its own colony of flagellates. It usually does this by eating the skins and gut linings shed by its elders when they moulted. Once an individual

becomes adult, of course, it no longer moults, so newly-hatched cockroaches have to stay close, not only to their parents but to their older but still juvenile siblings in order to acquire their digestive helpers. In short, cockroaches tend to live in family groups.

At some time or other, descendants of these early cockroaches began to form much larger communities, each one a gigantic family. When that happened, a new level of social living had appeared on the lands of the earth. These creatures were the early termites. Most of them are very much smaller than cockroaches and are only a few millimetres long. The evidence that they are related to cockroaches comes from many characteristics that are common to the primitive members of both groups. The patterns of the veins on their wings are very similar. So are the teeth on their jaws and the genitals of the females. Both lay their eggs in a similar fashion – in packets of a couple of dozen. Both eat cellulose as the major part of their diet and both digest it by the help of protozoan flagellates in their gut. And – most tellingly of all – those flagellates are closely related.

When this development happened is still a matter of speculation. The earliest undoubted termite fossil so far discovered is a single wing found in the rocks of Labrador which dates back to the middle Cretaceous, around 100 million years ago. Its vein pattern, however, is already quite advanced and expert opinion is that the origin of the group must lie very much earlier, perhaps even as far back as late Palaeozoic times around 250 million years ago.

By fifty million years ago, in the Eocene, these primitive termites were widespread in the world. Today only a single species of this group still survives. It is *Mastotermes darwiniensis* and it lives in tropical Australia and nests in the soil. A single community may number over a million. Its appetite is almost unbelievably wide. Wood is its staple diet but, like primitive cockroaches, it will eat an extraordinary variety of things and has, in fact, the most wide-ranging tastes of any living termite. It will tackle wool, horn, leather, rubber, the plastic coating of electric cables, even – it is said – billiard balls.

Colonies are founded by winged adults which fly out from nest holes in great numbers during certain seasons of the year. Males and females flutter through the skies away from their parental colonies and after travelling for as much as a kilometre or so, alight and shed their wings. Then they begin to explore the territory in which they have landed. The female takes the lead, followed enthusiastically by a male. Eventually she finds a crevice to her liking and settles in. They mate and the female produces her packet of eggs. When the young hatch, she feeds them with her spittle. These immature babies are small but active and in essentials

△
Winged, sexually mature termites come to ground and, having mated, shed their wings. Largely defenceless, they are immediately attacked by ants.

similar to their parents except that they lack wings. Like cock-roaches, they also imbibe secretions from her anus and so acquire an inoculation of the micro-organisms that are essential for their digestion.

Genetically, these young are both male and female but they have not yet developed their sex organs. Nor will most of them ever do so as long as they continue to receive food from their mother for the secretions she produces contain a hormone that inhibits sexual development. Instead of breeding themselves, these offspring devote their time to rearing the young that hatch from the eggs their mother continues to lay.

But this is by no means a permanent situation. The founding couple do not seem to have very long lives. When, for whatever reason, they disappear and the inhibitory hormones the female secreted no longer circulate in the colony, some of the young workers will develop sex organs and take over the task of produc-ing fertilised eggs. Sometimes this may happen in a part of the colony's workings that are some distance from the main nest and connected only by the passage-ways they built to reach it. Then a secondary nest forms with its own new king and queen.

This social malleability is retained by a group of families that although they are more advanced than *Mastotermes* are nonetheless known as the lower termites. Their young workers can also, if necessary, moult into sexually mature forms. What is more, they can even reverse the change and moult back to become sterile workers again. These lower termites also have flagellates in their guts to facilitate the digestion of cellulose. The newly hatched members of the colony acquire these not by eating the shed skins of their elders but by imbibing the secretions from the mouth or anus of their elder siblings.

In another group known as the higher termites, social relation-ships are more restricted. A queen, the founding female, is nearly always present. When fully developed, she is huge, a hundred times bigger than her original size. Pale, bloated, and thin-skinned

she lies in her own chamber, her soft body slowly rippling as the processes of egg production proceed within her. She may be as much as 14 centimetres long (5½ inches), so big that she cannot enter the passages that link her chamber to the rest of the nest.

Her mate, the king, lies alongside her. He is scarcely bigger than one of the workers but he alone will mate with her and he does so repeatedly throughout her life. She produces several thousand eggs a day. The individuals that hatch from them include both males and females. When they hatch they are miniatures of the adults. As they feed, so they grow and moult until finally they reach adult size. At certain times of the year, influenced doubtless by the queen's hormonal secretions circulating within the nest, tens of thousands of them will develop wings and genitalia and fly out of the nest in dense clouds to mate and found new colonies. Most however, will remain within the colony as sterile workers.

△
A queen termite lies in the royal chamber with her consort, the only fertile male in the colony, lying alongside. She is surrounded by workers who collect her eggs and ingest secretions from her body.

The workers are responsible for gathering the colony's food. In some wood-eating species they remain inside the wood in which the community has excavated its nest. In other species, they march out to collect it from the surrounding vegetation, dead or alive. All are blind but even so they dislike the desiccating heat of the sunshine, so if they have to venture outside, they usually do so at night and will often laboriously build temporary tunnels of mud to shield themselves. Within the nest, they have the responsibility of building extensions to the colony's accommodation. They must also keep the chambers and galleries clean. They feed the queen and collect the eggs as she produces them, carrying them away to special nursery chambers. The larvae, when they hatch, are by no means helpless. They are small versions of the workers, quite active and capable of feeding themselves, but they still need tending by workers acting as nursery maids.

Worker termites busy within the colony's fungus garden.
▽

When these communities are mature, which may take several years, a new kind of worker with a huge head starts to hatch from some of the queen's eggs. They are soldiers who have no duties other than to protect the community against intruders. In many species they have formidable jaws which are powered by enormous muscles in the swollen head. With these they are quite capable of slicing a tiny gash in human flesh. In other species the soldiers are armed in a different way. They have lost their jaws and in their place have developed a spout through which they can squirt a viscous liquid that entangles the body of any attacker. Exceptionally one family, the *Globotermes,* do not produce soldiers at all. Instead ordinary workers take on the responsibility. But they have a very effective defence all their own. They have become suicide bombers. When confronted by an enemy, they explode and disable their opponents by covering them with the sticky remains of their bodies.

△
Soldier termites on the march, ready to protect the food-gathering workers. These have spouts on their heads through which they can squirt an entangling liquid.

The most advanced species, among the higher termites, have in many instances dispensed with the assistance of flagellates in their guts. Instead, they maintain cultures of a much smaller kind of micro-organism, bacteria. These also help to turn cellulose into a digestible form. Some of this group of termites living in Africa and Asia – though none in the Americas – have an additional way of dealing with cellulose. They maintain gardens in their nests on which they cultivate a fungus. Each species of termite has its own species of fungus which grows in no other kind of nest. None of the species of fungi they cultivate has yet been found in the outside world.

The workers of these species, like those of so many others, live by industriously and often highly destructively, gnawing their way through dead wood. They fill their stomachs with it and return to special chambers in the nest in which their gardens grow. There they excrete the still-undigested wood fragments and mould them into a crumpled spongy pale-grey mass on which their fungus grows as a mat of threads. They then feed the resulting compost to the colony's developing larvae.

◇

These huge communities of termites need homes. Some species live in trees and use wood pulp mixed with saliva to construct great globes as much as a metre in diameter from which corridors will run into the wood which its inhabitants are currently consuming. Others live on the ground and build their nests using a mixture of earth, spittle and excrement that sets into an excellent and versatile building material. It is sometimes described as being as hard as concrete but that comparison is not really apt for although it is quite strong enough to build into huge towers it can nonetheless be easily broken by a shovel or a bush knife. But with this termites are able to construct some of the biggest of all animal-made dwellings.

Some may be low domes covering an area of ground as much as 30 metres in diameter. Others are towers with a series of roofs built one above the other like a Chinese pagoda which between them keep off the rain very effectively. The biggest contain many

tons of earth and are furnished with turrets that may rise as high as 6 metres (18 feet). Such huge mansions containing several million inhabitants must necessarily have mechanisms to ensure that the internal environment does not become intolerably hot or dry or lethally short of oxygen.

Different species have evolved surprisingly different ways of doing this. A particularly remarkable one is that built by *Amitermes meridionalis*, a species that lives in northern tropical Australia. Its architecture is simple. Understanding why it is so effective and why no other species of termite has adopted it, is less straightforward. These nests are shaped like huge chisel blades and stand clustered together and all facing in the same direction, like ordered ranks of immense tombstones in a crowded cemetery. In the dry season, they are surrounded by green grass even though

△
Soldier termites neither collect food nor tend the queen. Their sole responsibility is defence. In this species they are equipped with powerful jaws that can slice through skin.

the land elsewhere is parched and brown. In the wet season their land becomes a billabong, a shallow temporary lake.

Face on, they may measure up two metres (6 feet) across and over three metres (9 feet) high. In profile, they are barely half a metre at the base and only a centimetre or so across their blade-like top. Their crests all run north and south as though they were compass needles. This gives their makers their popular name of magnetic termite.

And the name is appropriate for they do indeed use the magnetism of the earth to guide them as they build. That has been proved by experimenters who buried an array of magnets around one of their colonies. These were sufficiently strong to distort the earth's magnetic field in this one small area. The termites, which continued to build within their nest, started to construct chambers that were skewed from their normal orientation by just the amount that the magnets had distorted the earth's field.

Building in this shape and aligning it north and south is a way of controlling the nest's internal temperature. In the morning the rising sun strikes the eastern face of the nest four-square so that the galleries immediately beneath the surface warm up rapidly after the cool of the night. As the sun climbs and continues on its journey from east to west, so its rays strike the tall near-vertical face more glancingly and therefore not so powerfully. By the time it is at its highest and almost directly overhead, this being the tropics, it strikes only the sharp top edge of the blade and its effect on the temperature within the nest is minimal.

But the day is not yet at its hottest. The temperature of the air is still rising and as the sun moves on and falls on the western side of the nest, the surface of the wall becomes so hot it is almost painful to the touch. But the eastern flank is now in shadow and the termites, most of which are still clustered beneath it within, are no longer being heated. Should the day be unusually cool, they may migrate to the western flank to catch the last warming rays of the sun, but for most of the time they remain near the eastern flank.

▷ ▷
A field of magnetic termites, near Darwin in northern tropical Australia. The better-drained area beyond their grassy territory, where trees are growing, is occupied by termites which build more conventionally shaped pyramidal mounds.

227

Groups of these termite hills are often surrounded only a few yards distant by those of other closely related species which instead of building blade shapes construct tall turreted castle-like dwellings. Why should this be? The answer seems to be connected with the seasonal flooding. In the heat of the day, most termites elsewhere in the world are able to retreat downwards to cool cellars below ground level. In the wet season however, magnetic termites cannot do this for the lower sections of their nest are flooded. But the simple yet ingenious shape of their homes enables them to deal with the problem in their own unique way.

But why should this solution be restricted to just this one species of Australian termite? The earth's magnetic field is gradually changing at a rate that varies according to where you are on the globe, as navigators and explorers who use magnetic compasses know well. Their charts specifically instruct them to apply differing annual variations to the direction of magnetic north in different parts of the world. The variation affecting the Australian continent however, is remarkably small. It is thus possible for *Amitermes,* guided by a relatively static magnetic field, to build with the same constant orientation for the century or so that it takes to construct such giant nests. Termites elsewhere, in places where the annual magnetic variation is greater, do not get sufficiently unvarying guidance to be able to do that.

The magnetic termites' homes are comparatively simple internally. Other termites, particularly those in Africa and Asia that cultivate fungi, have internal structures and overall arrangements that are much more complex. Termites have very specific environmental needs. The atmosphere must be relatively humid for they have very thin permeable skins and can easily dry out. The oxygen content of the air has to be maintained at proper levels if the inhabitants are not to suffocate. And the temperature must not vary extremely or rise too high.

All these environmental characteristics are kept at the most advantageous level entirely as a consequence of the nest's internal design and require no special actions from the occupants. The outside walls are massively thick, insulating to a considerable degree the living accommodation that lies between them. The air rises up the chimneys and flues, driven by the heat generated in the colony's fungus gardens together with that created by the daily exertions of the termites themselves. Some of the biggest nests are encircled by hollow buttresses rising on their outer flanks. The walls of these are made from a mortar with a porous texture so that gaseous exchange between the air outside and that inside can take place, carbon dioxide drifting out and oxygen filtering in. Most of the nests have deep cellars and in these the termites dig wells that may descend for 40 metres (120 feet) in order to reach the permanent water table. In some the cellar ceiling is covered by delicate earthen fins like the radiating gills on the underside of a giant mushroom, except that they have lacy lower margins. Each termite species has its own typical architecture but even within a species there can be local fashions or styles that may be appropriate to the area in ways we do not understand. One species, *Macrotermes bellicosus,* arranges its internal plan in one way in Nigeria and in a quite different way in Uganda. Yet all these different and often highly complex buildings are constructed in total darkness by tiny blind workers scarcely as long as your finger nail.

Blind, and in any case working in total darkness, a worker termite adds its little mud pellet to the colony's great construction.
▽

We can watch them at work. One will advance to the building site within the nest and with its antennae carefully examine the place where the next brick is needed. It then turns away so that the tip of its abdomen is above the place where the brick is to go and excretes a tiny drop of fluid from its anus. While in this position, it picks up a small pellet of moist mud in its jaws. Then it turns and places its contribution on top of the liquid drop which acts as cement and firmly presses the pellet into position with its head.

But how does each of these thousands of builders know how to place its brick so that it will be in accord with the overall design of that particular part of the nest? How do their cumulative actions result in the graceful arching vaults, the diffusing fins, the deep wells, the wide ventilating chimneys, the labyrinth of tiny intricate passageways and the spacious royal apartment in its correct position right in the heart of the building. And how, when the mating season comes, do they know to build special runways along which their winged sighted siblings can launch themselves into the outside world, or to perforate the top of their mound with hundreds of release holes and three minutes later reseal them after their fertile siblings have left. We have no idea.

◇

Some time after the termites established themselves, members of another group started on their own road towards becoming super-organisms. They were the wasps. But whereas all termites today live in large communities so that the intermediate stages that led to this condition must be largely a matter of speculation, wasps live at every level of sociality and include species that can illustrate the stages that have led to the development of this way of life.

Hunting wasps provide food for their young by laying their eggs on living meat – a caterpillar or a spider, a weevil or a fly. Not all, however, are as irresponsible as the species which parasitises the caterpillar of the Large Blue butterfly and then leaves it in the care of ants. Many hunting wasps go to great trouble to provide for their offspring themselves.

△
A female
Ammophila *wasp,*
having previously
dug her nest hole,
returns to stock it
with a caterpillar
that she has
paralysed with her
sting.

The technique probably began around 144 million years ago in the early Cretaceous and it is still practised by many hunting wasps. The female catches a caterpillar and paralyses it with her sting. She then starts to dig a short tunnel. Having completed that, she drags the caterpillar down into its tomb. Only then does she lay her egg on it. This method has one obvious disadvantage. The paralysed caterpillar must necessarily lie motionless and undefended while its sepulchre is being excavated and so is an easy target for thieves.

Other hunting wasps, however, show more premeditation. *Ammophila* digs her nursery chamber *before* she catches a caterpillar to put into it. That way she can take her time and dig a longer nest-tube without putting her capture at risk. Doing this, however, demands an additional talent. Having finished her tunnel and set off to hunt she has to be able to find it again when she returns with her prey. All nest-making insects have to be able to do this in one way or another, but whereas ground-living species can do it by following scent trails, flying insects need visual clues to guide them. So having finished her digging, *Ammophila* makes a number of orientation flights around her tunnel, noting the position of its entrance in relation to a prominent pebble, a small twig or whatever conspicuous landmark may lie nearby. That you can easily demonstrate if you are so unfeeling as to change the position of her pebble or twig, for if you do, she is quite baffled on her return and as like as not will eventually fly off and dig somewhere else.

    Ammophila is largely solitary. But other hunting wasps build chambers for their larvae close to one another. *Trigonopsis*, a Central American wasp, makes small pots out of mud and stocks them

▷
The beginning of sociality. Female Mischocyttarus *wasps build their cells together so ensuring that the eggs within are never unguarded. Panama.*

An American potter wasp finishes off her neat mud pot by adding an elegant out-turned rim, the mud of which is still dark and wet.
▽

with paralysed cockroaches. Up to four females that do so will construct their cells alongside one another. Each may make several pots, so work on the little cluster is more or less continuous with the females flying back and forth either with more mud for building more pots or more prey for provisioning them. Nesting in this way, they ensure that their cells are seldom if ever left unguarded. The females may occasionally assist one another in building their mud cells, but that is the limit of their cooperation. There is no division of labour between them and no difference in their physical appearance.

But only two small changes in this behaviour are needed to create a highly social life. First, one female in such a small colony should become sufficiently assertive to dominate her companions and lay her eggs in the cells that they construct and provision. And second, she should lay her first brood sufficiently early in the season for her young to reach maturity while she is still alive. If they devote their energies to raising their siblings who have exactly the same genetic make-up as themselves, then they will propagate their genes more effectively than if they were to fly away and dilute them with genes from a strange male. By doing these two things, wasps have developed colonies consisting of several thousand individuals that live in gigantic nests.

Such developments have taken place separately and independently at least four times in various families of wasps and the nests they build take many forms. In Peru, *Myschocyttarus* wasps build their individual cells one beneath another to form a long knobbly twig suspended beneath rocks. Most colonial wasps, however, join their six-sided cells into horizontal rafts known as combs. The parasol wasp, *Apoica*, another South American species that is nocturnal, constructs a huge single comb from chewed up vegetable fibres. This it hangs beneath branches and thatches with a layer of fibres and thin roots. There is no surrounding covering envelope and the several hundred members of the colony, which are particularly large and have startlingly white or yellow abdomens, spend the day hanging beneath it, clustered closely together.

▷ *The pyramidal paper nest of the social wasp* Polybia. *Here they are tapping the nest in synchrony as a warning of danger. Peru.*

▷ Apoica *wasps keep their horizontal comb continuously covered and guarded, with the members of the colony neatly positioned, head-outwards round the entire rim.*

Species with bigger colonies suspend a second raft beneath the first, attaching it by strong vertical rods. They usually enclose the whole with an enveloping wall which has a hole at the bottom to serve as an entrance. Remarkably, the builders of such multi-storeyed dwellings seem to be well aware that hanging additional combs puts a greater strain on the suspending rods at the top for as they enlarge the nest, so they thicken them to give them greater strength. Species of *Chartergus,* a genus of wasps found throughout the tropics, make smooth pale grey tubular cylinders 25 centimetres (10 inches) long. There is an entrance in the centre of the bottom surface that leads to a shaft running up the middle and gives access to the dozen or so horizontal combs, that rise tier upon tier above. *Parachartergus* does much the same but adds a long entrance spout at the bottom. *Polybia* uses very fine clay to mould a globular container with a ceramic-like surface that hangs from a broad strap around a branch and has a long vertical slit up the side as an entrance.

The common European wasp, *Vespula vulgaris*, chews up wood fibres to make a paper which it moulds into a vast globe. Sometimes it constructs this underground. The workers find a suitable vacant hole made perhaps by some other animal and steadily remove earth around the sides to turn it into a larger chamber as their nest increases in size. Sometimes they use accommodation that we unwittingly provide and create the monstrous growth we sometimes find squatting so menacingly in our lofts.

When flowering plants appeared, 120 million years ago, many worker wasps started to sip honey, although they continued to feed their larvae with animal tissue. Some of their relatives, however, accepted the bribes offered by the flowers as their sole diet and started to stock their cells not with paralysed caterpillars or spiders but with pollen and honey. These were the ancestral bees. The disadvantage of such a diet is that in many parts of the world it is only available for a short period of the year. Bees therefore have to labour intensively during this time, gathering as much nectar as they can, and then storing it in their nest in sufficient quantity to sustain themselves during the fallow part of the year.

So bees' nests containing cells filled with pollen and honey became particularly attractive to thieves and had to be guarded with great care. Many bees, like so many wasp species, were solitary and still are. But in time some found it advantageous to nest in groups and so ensure that their well-stocked larders were guarded continuously.

In spring, in Europe, a female bumble bee, waking from her winter hibernation and carrying within her the still-viable sperm received when she mated at the end of the previous summer, starts to fly up and down hedgerows looking for a suitable place to make her nest. She drones ponderously back and forth close to the ground. Sometimes she alights and explores the leaf litter before taking off again. Eventually she finds a small hole and pushes her way inside. Very often it is the abandoned nest made by a field mouse. If it is to her liking, she starts to enlarge the chamber, nibbling away at its walls, removing soil particles but leaving rootlets and other fibres so that it acquires a soft matted lining. She modifies the entrance, turning it into a narrow tube which sometimes can extend underground for several feet. Then she produces a building material that no wasp employs. Wax oozes from glands between the segments of her abdomen. It emerges as small flakes and with these she fashions a tiny cup. She fills it with pollen gathered from the spring flowers and on this she lays a dozen or so eggs. Then she seals the pot with a waxen cover. And to provide for her as yet unhatched young, she makes another small waxen pot nearby which she fills with honey. She then incubates these first eggs by sitting above them, keeping them warm with heat generated within her body

When the grubs appear, she provides each of them with a wax cell in which to live and feeds them assiduously with either a mixture of honey and pollen or pollen alone. Eventually they pupate and emerge as adult females. These, the first batch of her daughters, are usually rather small and they do not venture out of the nest. Instead they stay with their mother, the queen, and help her to expand their nest by digging away at the walls and building more wax cells. The queen forages for all. She also continues to lay. Her next batch of daughters are somewhat bigger and they start collecting food for the colony, allowing the queen to concentrate her energies on egg laying. These young females, however, are not totally incapable sexually. Some may indeed lay but if they do, the queen will eat their eggs. Her daughters must devote their energies to feeding their siblings and not their offspring. Eventually, there may be as many as 400 individuals in the nest.

In the late summer the queen starts to produce a different kind of egg. Until now, those she has laid have all been fertilised by sperm from the store she acquired during her mating flight and still retains within her body. All of these have been female. But the eggs of wasps and bees, as well as those of some beetles and cicadas, are also capable of developing without amalgamating with a male sperm. If they do so, they will develop into males. And that is what happens now. These young males differ slightly in colour and unlike their mother or sisters, they have no sting. Eventually, when the conditions are suitable, they will leave the nest and look for females.

But the queen also continues to lay fertilised eggs. The females that hatch from these tend to be slightly larger than their older sisters. There are good practical reasons why this should be so. The number of workers caring for the larvae is now higher than at any other time. Furthermore, the queen's output of eggs is declining. With fewer larvae developing and more workers to look after them, each of these late larvae is being very well fed. As a result, they are able to develop stores of fat within their bodies.

As autumn arrives, the colony goes into decline. Then both males and females leave the colony and look for mates. The workers and the old queen die, but the large young females that have now been mated seek holes and crevices in which to hibernate. There they are able to survive the winter, thanks to the fat stores in their bodies. They will be next year's queens.

◇

▷
A bumblebee colony in its early stages. The queen's daughters have been busy building cells in each of which the queen has laid an egg. Some, unsealed, they have filled with honey.

Bumble bee colonies may contain a few hundred individuals. Honey bees, however, live in communities that number tens of thousands. Like colonial wasps, they build six-sided cells which they join together to form combs. These are not horizontal, like wasp combs but hang vertically and are double-sided.

The bee that provides us with the bulk of our honey is native to Africa, the Middle East and Europe. People learned long ago how to provide it with hives, of wood, straw or pottery, so that they might more easily collect the honey. In the wild, however, the species nests in tree holes or rock clefts. A single colony may contain as many as 80,000 individuals and have several vertical combs, hanging one behind the other.

Unlike bumble bee colonies, these great assemblies do not die off with the coming of autumn and they may continue inhabiting the same location for many years. They also differ from bumble

△
Wild honey bees, nesting under a rock overhang in Brazil. Their combs, unlike those of wasps, hang vertically. Horizontal nests, open at the bottom like those made by wasps, could hardly hold liquid honey.

bees in that the queen, the males, and the workers are easily distinguished by their size, the queen being the biggest and the workers the smallest.

The queen is the only member of the honey bee colony to lay eggs. At the height of her productivity, she produces around fifteen hundred a day. For nearly all her life, she remains in the nest waited upon by her daughters who bring her food, regurgitating it from their stomachs and transferring it from their mouths to hers. In exchange, she gives them some of her spittle. This contains a liquid known as 'queen substance' or, perhaps more informatively, 'bee milk'. Her daughters take it with apparent eagerness but in doing so they condemn themselves to a life of drudgery, for bee milk contains a hormone that prevents the growth of their ovaries. They will remain forever sterile. They are the workers.

A worker's duties change as she matures. For the first three days after emerging from the cell in which she spent her larval life, she labours as a cleaner, removing any detritus or dead workers from the combs. For the following week, she is a nursemaid, feeding the developing larvae with a secretion from glands in her mouth, rich in vitamins, sugars and other important ingredients, which is

Worker bees exchange spittle with one another throughout their lives, so awareness of changes, such as those instigated by alterations in the queen's secretions, rapidly spreads throughout the colony.
▽

known as 'royal jelly.' These glands, however, cease to function after a further week and she has to change the diet of her charges. She feeds them now on pollen and honey which she draws from the storage cells. But now she develops a new set of different glands – the ones in the joints between the scales of her abdomen that produce flakes of wax. So she becomes a builder, labouring away to add more hexagonal cells to the combs, either to provide more storage for the colony's reserves of pollen and honey or new cells into each of which the queen will lay an egg. She then graduates to yet another job. She becomes a shelf-stacker in the colony's store, taking pollen and nectar from workers returning from foraging among flowers outside and packing it away in the new cells. When she is about three weeks old, she starts to spend her time close to the entrance of the nest, doing her best to repel any other creature that tries to enter, including bees from other colonies which do not smell right. Then, at last and for the first time, she ventures out of the nest and starts to collect nectar and pollen. And that task will occupy her for the last fortnight or so of her six-week life.

Like a wasp, she can remember the local landmarks, so having found food she has no difficulty in finding her way back to the nest. But she has an additional talent and an astonishing one that sets her apart from all other insects. When she arrives back in the hive with food she is able to tell her sisters where she found it.

If her source lies quite close to the nest, she makes her way through the throng that covers the comb and begins to dance excitedly. She travels around a small circle which has a diameter about twice the length of her body. Having done so in one direction, she turns round and goes back along the same track but in the opposite way. The richer the source of food she has found, the more frequently she changes direction. Several other workers trail behind her as she dances in the darkness. They sense her movements by feeling her with their antennae and from the excited high-pitched buzzing that she makes. They may even be able to learn what kind of flower she has visited from the faint perfume that still clings to her hairy body. Soon afterwards, stimulated by her performance, they themselves make their way out of the nest to try and find this new source of food.

If that source is more than about 25 metres (80 feet) distant, she must necessarily provide more detailed instructions if others are to find it. She walks up the comb in a straight line, waggling as she goes. Having gone a particular distance, she curves away to one side and returns to the beginning of her straight path, still waggling incessantly. Up she goes again but this time when she reaches the end of her straight path she curves away to the opposite side. The straight line that forms the junction between the two semi-circles is the most informative element in her performance. Its length indicates the distance of the food source away from the nest and its angling up the comb indicates the direction in which she flew to find it.

If the source can be reached by flying directly towards the sun, then she dances vertically. If it is off to the left, then the workers trailing behind her on the comb understand that they must fly away from the nest entrance at the same angle to the sun as her dance path is to the vertical. Remarkably, some reports say that she may continue her dance for hours in the darkness of the hive, altering the angle of her upwards path to keep it exactly the same as the angle of the sun outside at that particular moment.

As the workers labour away on their various tasks, the queen stalwartly continues hers, laying an egg in each of the new cells as they are constructed. But as she ages, her production of queen substance starts to diminish. And as the colony increases in size, so necessarily what she does produce has to be distributed among a greater number of workers. Each worker inevitably receives less and less, and as they do so, their behaviour changes. First they begin to build slightly larger cells. The queen treats these rather differently. The eggs she lays in them are not fertilised by sperm from the store in her body, so they in due course hatch into males – drones.

The workers then start to build even larger cells. These are not placed alongside other cells in the comb but are attached to the face of them, usually close to their lower edge. The eggs the queen lays in these have been fertilised and will therefore be female but, as they develop, they are treated very differently by the workers. They are not given any honey or pollen but are fed exclusively on royal jelly from the workers' mouth glands.

When the male larvae pupate and emerge as adults, they begin to leave the nest and gather with males from other colonies a short distance away from the nest. The workers now begin to treat the old queen rather differently. They nudge and barge her and do not feed her as attentively as they did before. Eventually, and for the first time in her life, she leaves the nest. A significant proportion of the workers go with her in a swarm. The whole group flies through the air in a dense cloud and alights on a tree, a rock or a building. Some of the workers now leave the swarm and start searching for a suitable place for a new nest.

When they find one, they return and inform the rest of the swarm of its location by waggle-dancing. Sometimes several will return with conflicting suggestions but the swarm as a whole appears to decide between the two by accepting the recommendations of the majority. It takes to the air again and disappears into the location discovered by the scouts – or provided by an attentive bee-keeper.

Meanwhile back in the old nest, the pupae in the large queen-cells are beginning to hatch. The first one to emerge signals to her unhatched sisters with a piping call. If one of them emerges while she is present, there will be a fight and one will die. The new virgin queen now leaves the nest and joins the awaiting males. She mates with one. It is a quick but violent engagement, which he does not survive. She then returns to the nest. Over the next three or four days she may make as many as three of these mating flights every day and so accumulate enough sperm in her storage pouch to last her lifetime. That completed she may fly away accompanied by a contingent of her worker sisters and found another colony. Alternatively, she may return to the colony in which she hatched and there destroy any other young queens in their cells that may still remain so that she may reign as its undisputed queen.

In many areas, the number of bee colonies is largely governed by the number of suitable holes. Bumble bees face the same limitation. The Asian honey bee, however, which is found as a number of very similar species from the foothills of the Himalayas to the tropical forests of Borneo, has adopted a bold way of overcoming this limitation. It builds its nest out in the open. That means, however, that it has to have a very effective way of defending its honey against thieves. And indeed it has.

The Asian honey bee, *Apis dorsata*, is a giant, the biggest of all bees. The workers measure up to 19 centimetres (nearly an inch) long and are armed with a sting that not only contains a very virulent poison but is also particularly long and quite capable of penetrating the standard protective suits normally used by bee-keepers. Each colony builds a single huge comb up to 3 metres (10 feet) across. In the Himalayas, they attach their comb to the tops of cliffs hundreds of feet high. In forests, they choose giant trees whose trunks rise smooth and unbranched for 20 or 30 metres (60 or 70 feet) and whose immense crowns emerge clear of the forest canopy. Such trees may carry several dozen separate colonies which hang like huge black dewlaps from its branches.

Each comb is covered at all times by a cloak of bees several individuals thick. They cling with their bodies neatly aligned facing upwards. At any one time, 70 percent of the members of the colony are on guard duty, protecting the honey, while the remainder are away foraging for more. It is not merely the attack of a single worker that a honey raider – human or animal – has to fear. When a worker bee stings, it simultaneously releases a pheromone that is detected within seconds by workers both from its own colony and from others nearby on the same tree. Guided by it, they fly immediately to its source and add their stings to that of the first so that the predator, within seconds, has to face a mass attack. Few can endure it.

Nonetheless, stinging is an expensive form of defence for the colony, for when a worker bee loses its sting, it dies. It is therefore better for a colony to give warnings of danger before an attack is

▷
Above: the huge hanging comb of the giant Asiatic honey bee. It is usually suspended from a branch so high that few ground-living honey-eaters are able to find it. A single tall tree, emerging above the general level of the canopy, may hold several dozen combs each belonging to a different colony.

Below: the giant honey bees, covering and protecting their comb, neatly align themselves head-up.

launched. The guardians on the comb do this with a coordinated movement of their bodies, producing an effect like that of a gust of wind passing over a field of grain or an enthusiastic football crowd supporting their team with a Mexican wave.

Even so, the colonies are not without predators. Small lizards – geckos and skinks – creep up the tree trunks and around the margins of the comb to lick honey from the cells. At night a species of death's head hawk moth pays regular visits to the colony. It hovers in front of the comb, its wings beating so fast they are a blur and then swiftly dives into the cloak of defending bees. It may even wade fearlessly through them, unmolested, before it finds the honey cells it seeks and plunges its proboscis into one of them. The moth can behave in this fearless way for it is disguised convincingly as a bee. It has a pheromone that exactly matches that produced by the bees themselves.

Birds too sometimes raid the colony. Bee-eaters catch workers in mid-air and carry them in their slender beaks to a perch where

△
Death's head hawkmoths stealing honey from a comb. They are able to enter a hive without being attacked by disguising themselves with a perfume very similar to that emitted by the bees themselves.

they vigorously wipe the bee against a branch so that the bee is made to discharge its sting. Only then will a bird swallow it. Honey buzzards come too. They seek not honey, as their name suggests, but the bee larvae. They may fly at a comb and rip off a section with their claws as they pass, or perch above it and tear off great pieces with their beak.

A colony of giant bees may contain several hundred thousand individuals. Living together in such numbers requires a high degree of social organisation. To prevent aerial collisions, departing bees take off upwards and those returning approach from below. Sanitation is also dealt with in an organised way. Young bees at the beginning of their adult lives spend all their time on the comb. In the afternoons, however, they leave in groups to defaecate in the forest. This avoids dropping clues on to the ground that might direct the attention of a terrestrial honey-hunter to the existence of combs high above. It also acquaints each worker with the surroundings of the nest so that it will be able to find its way back to the colony when it graduates to become a forager.

These sanitation flights spatter the forest vegetation with tiny yellow flecks, a phenomenon that was once noticed by U.S. military in Vietnam unfamiliar with the forest and caused their Department of Defence to accuse the U.S.S.R. of chemical or biological warfare.

Forest plants in bloom cannot be found all the year round, in many areas. In parts of Borneo, flowering is largely over by May and the bees must find new feeding grounds. They prepare to migrate. They fuel up, filling their stomachs with precious honey until their comb is completely empty. Then, with their queen, the whole colony sets off on a long migratory flight that may last weeks and have several intermediate stops. Exactly where individual colonies go has not yet been accurately charted and it is still not known how they find their way.

While they are away, wax moths lay their eggs in the empty combs. Their caterpillars, when they hatch, begin to eat the wax. Soon, their feastings, together with the heavy rains, demolish the combs. But when the flowering season returns to the forest, the bee colony reappears and starts to build a new comb on exactly the same tree, often exactly the same branch as it occupied a year earlier. It seems that the colony is able to recognise its own particular smell that still lingers in what little remains of the comb of the previous year. The behaviour is the more remarkable since the colony's workers are now of a later generation none of whom will have visited the site before.

◇

Another branch of the ancestral wasps followed a different course. Instead of becoming vegetarians with a taste for sugar like bees, they remained hunters. But they pursued their prey on the ground. There they had no use for wings. Indeed, wings might be a real impediment when chasing through the undergrowth. Accordingly the workers in these communities, who do all the hunting, shed them. These are the ants.

The most primitive known ants so far discovered lie within a piece of amber that was found in the rocks of New Jersey. Amber is fossilised resin. This particular lump oozed from a tree some ninety million years ago. Two small insects crawling over the tree trunk became trapped in it. As the resin continued to flow out, it enclosed these two creatures and preserved them in microscopic detail. Their lack of wings, the small size of their thorax and the existence of a pair of glands at the back of it, are sufficient to justify calling them ants. Yet they also had short mandibles and a sting that could be extruded and both those characters are typical of wasps. Accordingly the two insects were called *Sphecomyrma*, which in Greek approximates to wasp-ant.

Whether these ancient creatures were solitaries or members of a colony, we cannot be sure. It is certainly the case, however, that no ant today lives a lonely life comparable to that of a hunting wasp such as *Ammophila*. All live in communities.

Ant societies have the same basic structure as termite societies even though the two groups are only very distantly related. They are founded by a single female who is sexually mature and has wings. She mated during a courtship flight and so acquired a stock of sperm which she retains in a special pouch inside her body. The males who are also winged soon die after this flight but she finds a nest site – a crevice beneath the bark of a tree, a tiny crack in the ground. It need not be big, for soon she will have plenty of helpers to enlarge it for her. Sometimes she even seals herself in her tiny chamber.

Now she prepares to lay eggs. The muscles in her thorax which powered her wings begin to atrophy. She absorbs their substance and redirects it to nourish her ovaries and begins to lay eggs. Here ants differ importantly from termites. Whereas termites lay fertilised eggs which develop into both male and female individuals, the queen ant – like queen bees and queen wasps – can lay eggs that develop whether they are fertilised or not. Those that are not will become males. Those that are fertilised will develop into females. And that is what these first ones are.

In some of the simpler ant societies, the queen may now go out and hunt for food, bringing back prey which she cuts into fragments and feeds to her larvae. In other cases, she stays inside the nest and sustains the larvae with secretions from the glands in her mouth.

This first batch of daughters are often small compared with those who will arrive later since they have only been given minimal food. And they are all wingless, as their sisters to come will also be. Their mother continues to lay and her first-born daughters, who will never themselves lay eggs, now start work collecting food for their younger sisters. As the queen continues to lay so the colony grows until it numbers hundreds, thousands or in some cases millions of individuals, all of whom are sterile sisters.

At some stage, governed by the annual cycle or some particular atmospheric condition, the queen starts to lay eggs that have not been fertilised by sperm from her store. These accordingly develop into males. And they have wings. She also continues to lay fertilised eggs. The female larvae that develop from these are also particularly well fed, since the colony is at the height of its success and they too emerge from their pupal cases with wings.

The nest is now at its busiest and most crowded. Hundreds of winged individuals, both male and female, are scurrying along its galleries and chambers alongside the much greater numbers of wingless sterile workers. And then suddenly, usually at night, the winged individuals all become of one mind. The time has come for them to depart and like smoke billowing from a factory chimney they leave the nest swirling up into the air in thousands. The cue that starts this exodus is probably a climatic one, for all the colonies of one particular species in the area will be behaving in the same way. So males and females from different colonies meet. Once more, there are mid-air couplings. Most of the males will fail to find a mate for they greatly outnumber the females. But all, successful as well as unsuccessful, will soon die. No kings lie alongside their queens in an ant nest as they do among termites. The mated females go their own way. They alight, shed their wings, and the cycle begins all over again.

This is the basic pattern, though among the eight thousand or so different species of ant there are many variations in detail. The simplest of their societies include the biggest of all ant workers. It is well-called *Dinoponera gigantea*, the 'giant terrible-ant' and it lives in the forests of Brazil and the Guyanas. You may see one hunting through the leaf litter, moving swiftly and incessantly exploring its surroundings with its long antennae which, like all ant antennae, are elbowed with a joint in the middle. It is a slim, glossy black creature with formidably large toothed mandibles. And it may be as much as 33 millimetres (1.3 inches) long.

If you have Amerindian guides, they will treat it with as much caution as they would a venomous snake. Guyanese people call it either the 'four-sting' ant because they say that four stings from it are fatal, or the 'twenty-four hour ant' since even a single sting will totally incapacitate you for at least a day, during which the severest pain spreads throughout your body. That may then be followed by a day of delirium and near-total paralysis before a victim recovers.

The nests made by these monsters are no more than a few galleries excavated in rotten wood or dug just beneath the surface of the ground and their colonies number no more than a few hundred. Although the queen is the only member of the community to lay eggs, she is only marginally larger than her working sterile daughters. Indeed, in some species she herself may share the labour of hunting. The prey is simply cut into fragments and then dropped beside the larvae which jostle and barge one another in collecting it and then chew it up for themselves.

Another similar family of hunters lives in Australia. There they are known as bulldog or jumping ants, for when disturbed they leap into the air. They habitually hold their jaws wide open. Should their jaws touch something they immediately snap shut and with such force that if the object is a smooth grain of gravel the jaws slip off and the ant itself is hurled into the air. These jaws are so large that the biggest species are longer than a *Dinoponera* worker, but the bulldog ant is slimmer and slightly smaller in terms of bulk.

In Africa another member of this group of ants, *Megaponera*, lives in colonies of several hundred. They are known as Matabele

An Australian bulldog ant drags its prey back to the nest. The adults feed primarily on nectar, but the larvae are carnivorous and eat insects gathered by workers. This ant's poison is potentially lethal to humans.
▽

ants, after the southern African people who, in days gone by, were famous and aggressive warriors much feared by their neighbours. The ants live entirely by raiding termites. Early on most mornings, several hundred of them emerge from their underground nest. The majority are large and formidably armed with great shear-like jaws that they carry pointing vertically downwards. Among them run minors, similar in form but only about half their companions' length. The column marches, half a dozen or so abreast, following a scent trail that had been laid down by a solitary scout earlier that morning. If you draw a line in the earth with your foot just ahead of them, you will break that trail. The column mills around in confusion for a few seconds but very soon one of them will have identified the trail ahead and once again they march on.

As they go, they make a rustling chorus of sound. They produce it by rapidly jerking their abdomens up and down, so rubbing a file-like patch of ridges on the front against a spike in the waist. They may march for a hundred metres or so until eventually they reach their target, a termite hill. Its entrances are guarded by ranks of soldier termites. Each is at least as big as the larger Matabele warriors. They have huge heads, heavily armoured with a helmet of shining polished chitin and filled with muscle to power their great jaws. But they are no match for the Matabeles. The ants seize them, sometimes by one of termite's jaws, sometimes by the back of the neck. Holding its victim thus, the ant brings forward its abdomen and jabs its sting into the only vulnerable part on the termite's armoured head, its mouth, just beneath the jaws, and injects a poison directly into the termite's brain. The effect is immediate. The termite staggers about, blindly snapping its jaws and collapses. Other Matabeles join in and sting it yet again, this time in its soft abdomen. Several of the raiders now start to lug the still twitching body down the passages to the entrance and dump it with others in a pile outside.

With the soldier termites overpowered, the termite nest is defenceless. The Matabeles race deeper inside to find the soft-bodied workers and slaughter them wholesale. The carnage

continues for a quarter of an hour or so and the piles of the slaugh-tered grow. The minor Matabele workers now take on the job of porterage. They gather around the dumps and with abdomens thrust forward under their thorax, they push the dead termite bodies into their open jaws. Some may gather up as many as half a dozen. Then the column sets off on their return journey. And once again there is a chorus of sound. This time it is noticeably different in its rhythms from the marching chants they made when they set out. Nor is it a uniform undifferentiated chorus. Listening to them with the help of a miniature microphone lying alongside their path, it is possible to pick out the sound of a single individual singing its own song against the background chorus, as it approaches and passes by on its triumphant journey back to the nest.

◇

The wood ants of the genus *Formica* that live in European conifer-ous forests, are also hunters. Their workers are rather smaller, a mere 6-9 millimetres (less than half an inch) long, but their com-munities are much bigger, numbering several hundred thousand. The nests they build are huge – great piles of pine needles that may stand as tall as a man.

They are abundant in Switzerland's Jura mountains. There in winter, the snow may lie two metres (six feet) deep, totally cover-ing the nests. The temperature at night may fall to 10° Centigrade below zero. Inside a nest at this time, there is no activity of any kind. Most of the workers lie in the galleries in the very founda-tions of the great pile, close to the ground surface. They are totally motionless. Their stomachs are full of honeydew that they col-lected from aphids at the end of last summer. But they have not yet digested it. They are saving it to serve as the start–up fuel when the winter at last comes to an end.

Spring arrives and the snow melts. As the surface of the nest reappears above the snow, its dark colour absorbs the heat and its temperature quickly rises. Those workers that are lying nearer to the surface of the nest slowly warm into life and make their way, moving awkwardly and stiffly, into the outside world. There they sun-bathe. As soon as they are fully warmed, they return to the

deeper part of the nest, moving with a new speed and vigour and circulate among the other still dormant workers, awakening them with their warmth. So more and more of the colony's inhabitants climb up to the surface to sun bathe. After a couple of weeks the entire colony has come to life. Work begins on the surface of the nest, repairing any damage that may have been caused during the winter. Working parties set out on the foraging trails, looking for prey. The wood ants are back in business.

They hunt, collecting insects from great distances around the nest and bringing their victims back to the nest to share with those workers that have been busy with domestic duties and to feed to the developing young. They also collect flakes of resin from the pine trees which they distribute throughout the nest. What purpose the resin serves is not certain. Individual ants appear to lick the surface of the flakes, but they do not eat it in any quantity. It seems that its pungent scent helps to fumigate the interior of the nest and keep it free from fungal and other infections.

Should you disturb the nest now, the workers will swarm out in a busy angry crowd, milling about all over the nest surface. They do not have stings. They repel intruders in another way. They raise their abdomens and squirt jets of formic acid in the air from glands in the end of their abdomens. This is quite enough to deter most insect invaders. Paradoxically, however, it actually attracts some of the forest birds. Jays, starlings and crows in particular fly down to the nest surface and deliberately stimulate the ants to squirt.

△
Wood ants swarming over the surface of their great nest repel intruders by squirting jets of acrid formic acid into the air.

As the insects do so, the birds crouch and erect their feathers so that the angry ants swarm all over them, discharging acid. In doing this, the birds rid themselves of any fleas, lice or other skin parasites lodged in their feathers. But the birds seem to enjoy the process for its own sake, quite apart from its sanitary effects. They cock their heads, sometimes with eyes closed as though lost in the delights of the stimulation the formic acid brings to their skins.

One species of wood ant, *Formica sanguinea*, collects more than prey. They regularly invade the nests of a related species, the brown wood ant, *Formica fusca*, and steal the pupae, seizing them in their jaws and carrying them back to their own colony. When these alien pupae hatch, they labour without protest in the red ant's nest, undertaking all kinds of domestic duties as if it were their own proper home. Since they are sterile, they cannot affect the long-term genetic future of the colony, which has now, without putting any additional demands on the egg-laying queen, acquired a population of slaves.

Another species, *Formica lugubris*, practises a different form of piracy. When sexually mature females fly out of their nest and are fertilised, one of them may land and enter a brown wood ant's nest. She makes her way through the workers, who apparently offer little resistance, finds the brown ant queen and stings her to death. Then she takes her place and starts to lay eggs. These are cared for by the brown ant workers as if they were those of their own species. Since there is now no brown ant queen, the original brown workers are not replaced as they die, so before long the whole colony has become a red ant community.

Red wood ant colonies can flourish without slaves, but other slave-making species cannot. An Amazonian species, *Polyergus rufescens*, has become totally dependent upon them. Its workers have huge sabre-shaped mandibles which make them very powerful hunters. But they are so big that their owners cannot feed for themselves. They therefore are compelled to raid *Formica* colonies to collect slaves who will perform this service for them.

◇

The most feared of all these ant raiders, however, are the army ants. Africa and South America each have their own families of army species – raiders who spend a significant part of their lives on the march. Those in Africa are known as drivers. When on the march, they form columns several inches wide, ribbons of running, tumbling, racing insects. Those in the centre are the swiftest, the small workers of the colony. On the outer margins of the columns march another kind of sterile female worker, the soldiers. They are much larger and have huge heads equipped with

▷
The massive jaws of a bulldog ant worker. Brazil.

sword-like mandibles. Some act as guards, lining the route facing outwards, ferocious jaws agape. In places where the column has to cross an open patch, the soldiers may link legs and form a living lacy roof to shield the workers beneath from the sunshine. It is not the light they dislike, for all army ants are blind. It is the heat.

At the head of the column, the workers are hunting. If they find prey – a scorpion, a grasshopper, a nestling bird, a millipede – they swarm all over it, sinking their mandibles into it, killing it and finally cutting it to pieces to send back to the main mass of the community who may be camped many metres away.

African villagers do not necessarily regard these marauders as pests. If they march into a village house, they will rid it of all kinds of unwanted lodgers – beetles, fleas, lice, cockroaches, even mice and rats. If such creatures do not flee, they will be cut to pieces. Animals that are tethered have little chance. Even horses and dogs can be so badly stung by the ants swarming all over them that they may die from shock.

Once a column entered a hut containing several dozen snakes – Gaboon vipers, spitting cobras, pythons – that we had collected for a zoo. Each was safely enclosed within a box with a metal gauze top. A watchman had been stationed beside the hut, armed with a can of kerosene and instructions to pour it in a line across the path of any approaching column and then set it alight. That, we thought, would deflect any drivers. Unluckily, the watchman fell asleep and a driver column marched in beneath the door and invaded all the boxes. By the time we discovered what had happened, every snake was being attacked. Each had to be taken from its box and held down outstretched with one person holding its head and another its tail while a third had the job of picking off the soldier ants that had fastened their jaws between the snakes' scales. Some survived. Most died from the shock.

A swarm of drivers may contain more than three million individuals. When stationary, the army builds itself a simple galleried nest just beneath the surface of the ground. But such an immense number of mouths demand a great deal of food. No neighbourhood can sustain them indefinitely. After three or four weeks, the land around the nest has been largely stripped of prey and the army decides to move on. The march usually begins at night and may continue for two or three days without pause. During that time the whole army may have travelled several hundred metres before they decide to settle and start hunting again.

◇

Army ants bivouacked beneath a boulder in the Costa Rican forest. Soldiers, clinging to one another, hang in a protective veil around the workers and the queen resting deeper within the huge mass.
▽

In the New World, from Brazil in the south to Mexico in the north, lives another unrelated group of ants that have also adopted the marching, pillaging life of an army. Their habits are slightly different from the African drivers for they spend even more of their time travelling and do so to a more predictable schedule. When camped, they form a bivouac, a vibrating tangled mass of half a million insects that measures a metre across, assembled beneath a boulder, between the buttresses of a forest tree or up on one of its branches. Somewhere inside, in one of the chambers or

galleries formed by the linked legs of the workers, lies the queen. She is considerably larger than any other individual with a bloated abdomen that at this time contains some 60,000 unlaid eggs.

Every morning, just before dawn, a raiding party of workers and soldiers sets out to collect the army's daily rations. Their needs are not as great as they will become for the majority remain in the bivouac and are not expending very much energy. The raiders march in a broad black column that snakes out from the bivouac, running over leaves, beneath logs and around boulders. In most species, these columns branch to form several smaller ones. In others they fan out to form a front that may be fifteen metres across. The workers in the vanguard advance a few centimetres into new territory, laying down a trail of odour from the tips of their abdomens. But then they double back and their place is taken by others, hot on their heels, following the odour trail. If they encounter animal prey, they seize it, swarm all over it, sub–duing it with their bites. Smaller creatures are killed quickly and

Army ants on the march scurry along in a broad ribbon-like column, snaking its way across the forest floor.
▽

△
*The army has
overwhelmed a
caterpillar and stung
it to death. Now the
workers co-operate to
transport the body
back to their
bivouac.*

then carried back. Larger ones, such as a scorpion or a millipede, are butchered on the spot and taken back to the bivouac in fragments, past other eager hunters racing up to take their turn at the front.

Back in camp, the booty is chewed and absorbed by more workers and then its substance is distributed throughout the bivouac by a continual exchange of spittle between individuals.

A few days after the establishment of the bivouac, the queen begins to lay in a sustained bout of extraordinary activity. In only a few days she produces up to 300,000 eggs. Each is taken away and cared for by the workers. When they hatch, the workers feed the young larvae and keep them clean. There are other young in the bivouac too. These hatched several weeks earlier during a previous stationary phase and are now small brown capsules, pupae. After about three weeks, more workers emerge from them too, although for the first few days these are rather weak and paler in colour than their dark brown older sisters. The bivouac is now at its

267

most populous, since it contains a whole new set of both workers and larvae. There are many more mouths to be fed. Not only that, but food has become increasingly scarce since the army has been hunting through the area for three weeks. It is time to move on.

One evening, some of the workers do not return to the bivouac. Instead they settle in a new place altogether. Others join them. Eventually the whole community is on the move streaming out and following the odour trail to the new encampment. The queen is among the last to leave. Her body, having got rid of so many eggs, is not as swollen as it was a week or so earlier and she is comparatively slim and much more mobile. Even so, she is surrounded by an excited retinue of workers and soldiers that cluster around her and even run over her as she makes her way down the trail to her new quarters. But she will only stay there for a day. The next night the whole army will move again and will continue

A soldier army ant stands on guard, its huge jaws agape, while workers beneath ferry the vulnerable larvae to a new bivouac. ▽

to do so every night for three weeks. In that time the eggs she laid during the stationary phase develop into larvae. When these begin to turn into pupae, the migration stops and the army settles into a bivouac that will be occupied for another three week spell. But by then they will be in country where they have not hunted for some time and there will be adequate food to fuel the queen's next bout of egg-laying.

◇

All ants did not remain hunters. Just as some descendants of the early wasps developed a taste for honey and gave rise to bees, so some ants also discovered its value as food and became honey ants. That development occurred independently among several different ant families and in several different places – in Africa, Mexico, New Guinea and Australia. All live in deserts and all have to endure long dry periods when their favoured food is no longer available. So all have to have a method of storing it and all have adopted the same bizarre solution. They force-feed a particular caste of workers until their abdomens swell to the size of small grapes. The chitinous membrane between the abdominal segments is stretched so tightly that it becomes transparent and the plates that once protected the abdomen are left isolated as pairs of tiny brown scales.

The Australian honeypot ant, *Camponotus*, is smaller than the average ant. You need sharp eyes to spot one scurrying over the desert sands or in the sparse eucalyptus scrub. Your eyes must be sharper still if you are to distinguish it from other small species that live in such deserts. One of these honey-collectors is plain brown. Another, a little easier to recognise, is black with two or three yellow stripes across its tiny abdomen. Most do their foraging during the night when it is cool. They collect nectar not just from flowers but from galls, from the secretions of aphids and from glands of unknown function that occur on the leaves of some species of eucalypts.

If you follow the workers on their return journey, they will lead you to a small hole, only a couple of centimetres in diameter

that descends vertically into the ground. It takes a lot of hard digging to follow it down for it may continue for a couple of metres before it branches. But it is worth the effort for at last you reach galleries that extend horizontally from the main shaft. Inside them hang row upon row of amber-coloured translucent globes, suspended by the bloated worker's tiny front legs. A single nest may contain a hundred or so. If you place one between your teeth, the skin will break under the slightest pressure and warm liquid honey spurts into your mouth. The desert has few greater delicacies.

△

Australian honeypot ants with their abdomens swollen with liquid honey hang from the roof of their underground gallery. A normal worker visits them for a drink.

Some ants have abandoned the habits of their early hunting ancestors and become true vegetarians. Indeed, those belonging to a
group known as harvesters were once believed to be active farmers for it is quite common to see particularly dense growths of just
one or two species of desert grasses surrounding their nest holes.
They thus have been praised from Biblical times onwards for their
industry and foresight. It is indeed true that these plants have
sprouted from seeds that were gathered from the surrounding
desert and placed there by the ants.

But these are not deliberately planted fields. On the contrary,
they are waste tips. The ants industriously collected seeds from as
much as a dozen metres away from their nest entrance and took
them down to their storage areas. But later, as they sorted through
their granaries, they discovered they had occasionally made mis-
takes. Some of the seeds are of a kind they dislike, so these they
take back to the surface and discard around the entrance where in
due course they germinate. The seeds they keep, however, do not
develop for the ants bite off the part of the seed from which the
first root will sprout to ensure that none will do so and thus, as far

*Harvester ants
gather seeds from the
desert sands and
hurry back to store
them in their nest.*
▽

as the ants are concerned, become inedible. So harvester ants are not farmers but gleaners.

In the cooler parts of the day, early morning and evening, harvester scouts emerge from the nest holes and set off along established trails to explore their surroundings. If a worker finds a single seed, it picks it up and takes it straight back to the nest, guiding itself by visual landmarks and the position of the sun. If, however, it discovers a group of seeds that may have recently fallen from the parent plant in some numbers, the ant will pick up one, which is all it can carry and as it hurries back it will lay a trail of odour from the end of its abdomen. This tells other ants in the nest that there are more seeds to be gathered and a group of them will hurry off to collect them before competitors do so.

Each colony of harvesters remains on its own territory and usually resists crossing the frontier into land which carries scent trails of a neighbouring colony. As the paths divide and divide again, they may mesh with others coming from another colony and run almost alongside. Yet each will avoid crossing the other.

But while colonies of the same species avoid interfering with one another, there is a no-holds-barred rivalry with others belonging to different species. In the Mojave Desert in the southwest of the United States, an aggressive and energetic harvester, a species of *Pogonomyrmex*, scours the desert with such efficiency that it seldom leaves seeds for others. Another ant, a species of *Aphaenogaster*, has discovered a way of thwarting *Pogonomyrmex* sufficiently to enable it to live in the same territory. Just before dawn, *Aphaenogaster* workers set out to visit the nest holes of *Pogonomyrmex* in the area and rapidly block their entrances with grains of gravel. An *Aphaenogaster* working party can achieve this in a relatively short time. The entrance hole, however, is so small that only a single *Pogonomyrmex* can work at clearing the barricade from beneath and it takes a significant time for it to clear the blockage sufficiently to enable a harvesting party to set out. By which time *Aphaenogaster* will have been able to gather a fair ration of seeds.

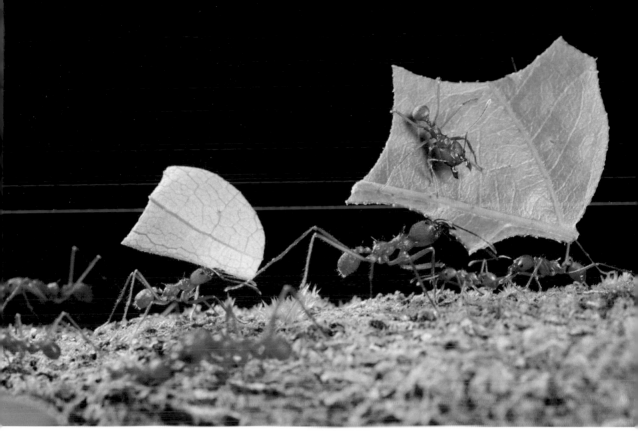

Leaf-cutter ants might also seem at first sight to be straightforward vegetarians, for processions of them carrying neatly sawn discs of leaves are a common sight in the tropics of the New World. But the leaves are not for eating. They, in a remarkable parallel to the development made by termites in Africa, cultivate gardens deep inside their nests. The crop they produce, however, is slightly different.

Their nests are gigantic, vast areas of naked earth, tens of metres across and dotted with as many as a thousand entrance holes. They lead to chambers which may be 6 metres (18 feet) below the surface. The whole nest can measure up to 3,900 cubic metres, the volume of a fair-sized village hall. Above ground during the day, there is seldom much activity to be seen. A line of workers may be marching out of one of the holes, each carrying a tiny brown fragment of faeces or perhaps the dead body of a worker. Nest cleaning is in progress. Elsewhere perhaps, a straggly broken line of workers carrying leaf segments is descending into another hole.

It is at night that the colony is most visibly active. Then columns of workers marching half a dozen abreast, come running out of the entrances. To begin with they travel along pathways several inches wide and worn smooth by the tread of a million minuscule feet. They follow increasingly tenuous tracks to the tree that is providing them with their current crop. Whole branches are covered with ant workers industriously scissoring out semicircular segments of leaves. As each makes the final cut, it tucks its jaws downwards, grips the segment firmly and then lifts its head so that the leaf is held over its back. Then it starts the long journey – a hundred metres, maybe – back to the nest. Other workers, unladen and on the outward journey, swerve to one side as the bearers stagger unsteadily onwards.

Inside the nest, they carry the leaf downwards along the smooth-sided tunnels to special chambers that contain a grey crumbling honeycomb. This is the fungus garden. As among termite gardeners, the fungus itself is a species that grows only in the nests of this particular kind of ant. The workers lay down their burdens and lick them clean. The segments are cut up still further

▷
Worker leaf-cutters labouring in the fungus gardens thrust leaf fragments into the tangle of fungus threads. A large-jawed soldier has also strayed into the garden.

A leaf-cutter ant grips a segment with two of its legs as it makes the final cut, to ensure that it does not drop it.
▽

with the assistance of other workers who are labouring on the fungus combs until they are reduced to fragments only one or two millimetres across. Their edges are then chewed to make them soft and moist and given a small drop of yellow excreted liquid. Then each is carefully inserted into the mat of fungal threads.

As the filaments of fungus grow they develop swollen knobs at the ends that are rich in protein. These the workers nibble and ingest and then use to feed the colony's developing larvae. They themselves, it seems, get most of their sustenance from the plant sap exuding from the leaf segments.

All the queen's daughters are not the same. As among the higher termites, they are differentiated into castes. The workers, who have the primary responsibility for cutting and transporting leaf-segments are about a centimetre (a third of an inch) in length. There are others, however, known as the minima, who are scarcely more than a quarter of this size. These midgets work almost entirely within the nest, labouring in the fungus gardens, caring for the larvae and keeping the nest clean. Occasionally, however, they accompany their bigger siblings to the tree where leaves are being gathered and travel back riding shot-gun on a leaf segment being carried by a big worker. Their responsibility then is to ward off any of the parasitic flies that might try to lay their eggs on the neck of their mount.

The biggest of all, apart from the queen, act as soldiers. They can be two centimetres (about ¾ inch) long and have swollen glossy brown heads and sabre-shaped mandibles. A couple of thumps on the ground with your fist is enough to bring them racing out from the nest entrances with their jaws agape ready to attack intruders.

In all, a mature colony of leaf-cutters may contain as many as eight million inhabitants. It is the biggest of all ant communities. No forest animal can seriously injure it. Neither can human beings do so, short of instituting a massive programme of poisoning or blowing up the nest with a substantial charge of explosives. The great super-colony lurks in the forest, a powerful perpetually hungry presence that is almost invulnerable.

A leaf-cutter ant community dominates all life around it as effectively as any plant-eating mammal. It consumes up to 17 percent of all the leaf production in the forest. In Death Valley, the population of harvester ants has a biomass equal to that of all the rodents living there. Termite colonies on an African savannah eat as much of the pasture as a large herd of antelopes. A community of army ants can weigh as much as 20 kilograms, and so match the weight of a clouded leopard. All these insects have formed closely coordinated societies that can properly be regarded as single entities comparable to the more visible mammals and in some ways more invulnerable than they are. They are the ultimate in insect evolution – so far.

The invertebrates were the first colonisers of the lands of this planet. They established the first of its ecosystems. Were mammals to disappear from the earth, the forests and their myriad invertebrate inhabitants would continue to flourish as they have done for more than four hundred million years. But were those invertebrates to disappear, the earth's ecosystems would collapse. A high proportion of plants would no longer be fertilised and would die out. Many amphibians, reptiles, birds and mammals that have depended upon a diet of insects ever since they evolved would starve. Dung and corpses would no longer be recycled. The creatures that perform all these essential services are still all around us unnoticed in the undergrowth. Many are within a few inches of our feet wherever we tread on earth, usually un-regarded.

We would do well to remember them.

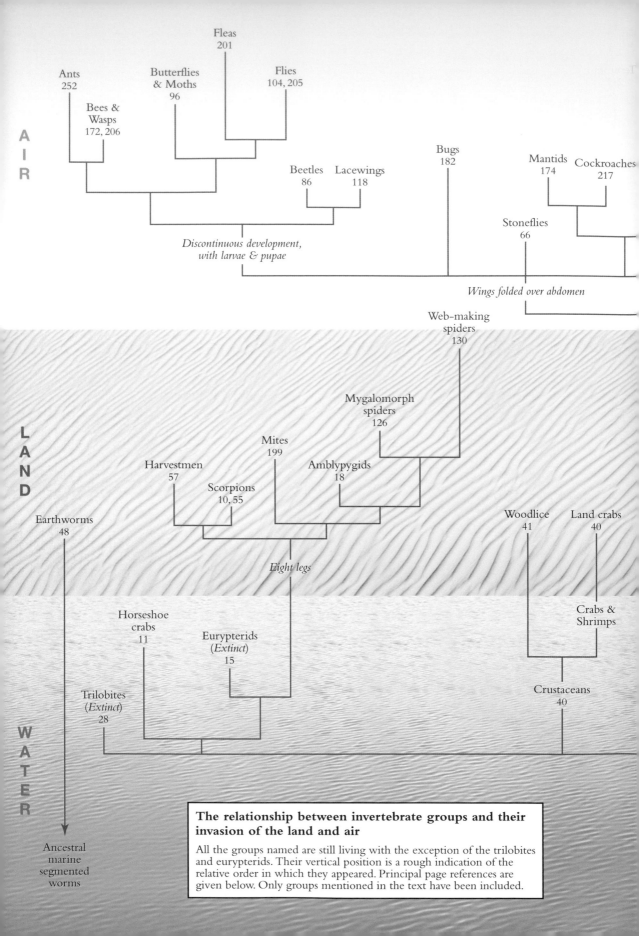

AIR

Fleas
201

Ants
252

Butterflies
& Moths
96

Flies
104, 205

Bees &
Wasps
172, 206

Beetles
86

Lacewings
118

Bugs
182

Mantids
174

Cockroaches
217

Stoneflies
66

*Discontinuous development,
with larvae & pupae*

Wings folded over abdomen

LAND

Web-making
spiders
130

Mygalomorph
spiders
126

Mites
199

Harvestmen
57

Amblypygids
18

Scorpions
10, 55

Earthworms
48

Woodlice
41

Land crabs
40

Eight legs

WATER

Horseshoe
crabs
11

Eurypterids
(*Extinct*)
15

Crabs &
Shrimps

Trilobites
(*Extinct*)
28

Crustaceans
40

Ancestral
marine
segmented
worms

**The relationship between invertebrate groups and their
invasion of the land and air**

All the groups named are still living with the exception of the trilobites
and eurypterids. Their vertical position is a rough indication of the
relative order in which they appeared. Principal page references are
given below. Only groups mentioned in the text have been included.

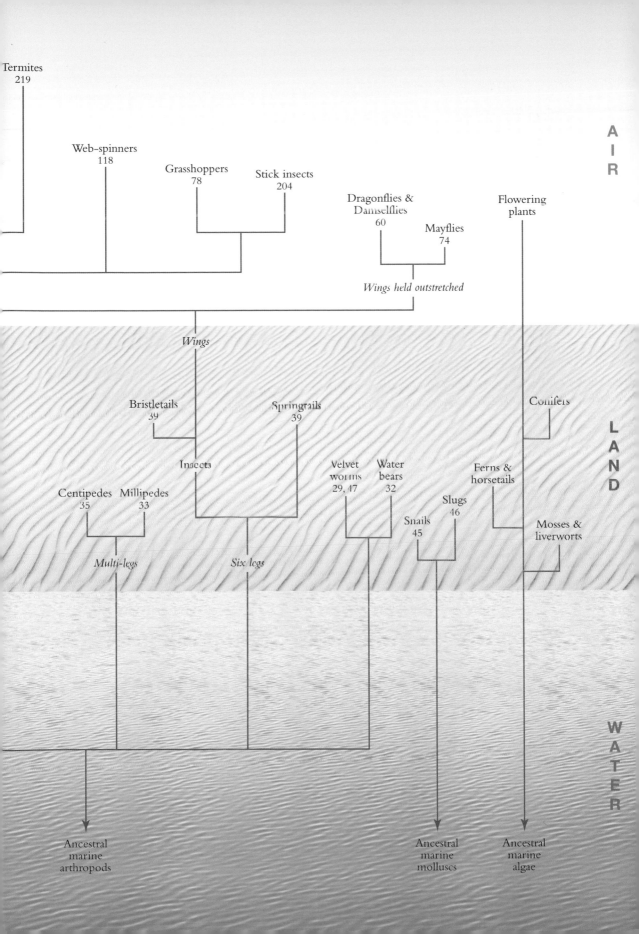

Acknowledgments

This book is one half of a single enterprise that tried to simplify the vast complexity of the world of terrestrial invertebrates. The other half was a series of television programmes filmed at the same time as the book was being written. Not surprisingly, one half greatly influenced the other. My initial film scripts were merely the base for vigorous debates about how each programme should be structured and what species should be featured. The resulting outlines were then inevitably modified by what species we found in the field, and sometimes even by the weather when we were looking for them.

Many people therefore contributed to this book and influenced its contents. Whenever we selected a species we were able to find a scientist who had spent years of his or her life studying it, who was able to show us where we might find the creature in question and who helped us to interpret what our cameras subsequently revealed. The debt that all of us working on the programmes owe to such experts – and there were nearly a hundred of them in all – is huge.

The film cameramen taught me a great deal about the creatures they spent so much time in observing with great concentration. The directors of each programme greatly influenced the chapters based on their films, improving the structure of the text as they did that of the programme. The researchers again and again produced ideas and details that astonished us all and which were incorporated into both the programmes and these pages.

The members of the television team are named opposite. There is not one of them to whom I am not indebted – and grateful for the great pleasure that comes from working with such a team for nearly three years.

Two people in particular have been invaluable in preventing me from falling into error – Alex Freeman who combined that formidable task with researching for the programme and George McGavin, Assistant Curator of the Hope Entomological Collections at the Oxford Museum of Natural History. I am extremely grateful to them both.

LIFE IN THE UNDERGROWTH

The Television Team

Executive Producer
Mike Salisbury

Producers
Bridget Appleby
Peter Bassett
Stephen Dunleavy
Andrew Murray

Assistant Producer
Tim Green

Researchers
Alex Freeman
Simon Williams

Production Manager
Ruth Flowers

Production Co-ordinators
Loula Charalambous
Lea Aldridge

Production Team Assistant
Vicki Hinks

Colourist
Jonathan Prosser

Picture Logging
Emma Napper

Music
David Poore
Ben Salisbury

Graphic Design
Mick Connaire
3D Animation – Andy Power

Photography
Tony Allen
Luke Barnett
Keith Brust
Rod Clarke
Martin Dohm
Steve Downer
Kevin Flay
Graham Hatherley
Nick Hayward
Richard Kirby
Alastair MacEwen
Peter Nearhos
Mark Payne-Gill
David Rasmussen
Tim Shepherd
Sinclair Stammers
Gavin Thurston

Additional Photography
John Brown
Rebecca Hosking
David Spears

Sound Recordists
Graham Ross
Chris Watson
Andrew Yarme

Film Editing
Tim Coope
Jo Payne
Andrew Mort
Bobby Sheikh
Vincent Wright

Dubbing Editors
Kate Hopkins
Paul Cowgill
Angela Groves
Paul Fisher

Dubbing Mixers
Pete Davies
Martyn Harries
Steve Williams

Online Editors
Steve Olive
Chas Francis

Series Scientific Consultant
George McGavin

The publishers would like to record their thanks to, and appreciation of,
Miriam Hyman
who researched the photograph that appears on the front cover of this book.
Miriam tragically died in the terrorist bombings in London on 7 July 2005.

Sources of Photographs

151 Oxford Scientific Films
(Densey Clyne Productions)

152 FLPA (Reg Morrison/Auscape)

153 Nature Picture Library
(Ken Preston–Mafham)

155 Natural Visions
(Francesco Tomasinelli)

156 Premaphotos (Ken Preston–Mafham)

157 Premaphotos (Ken Preston–Mafham)

159 Premaphotos (Ken Preston–Mafham)

160 *above* Nature Picture Library
(Ken Preston–Mafham)

160 *below* Nature Picture Library
(Ken Preston–Mafham)

162 *above* Oxford Scientific Films
(Alastair Shay)

162 *below* BIOS (Daniel Heuclin)

164 Nature Picture Library
(William Osborn)

165 Oxford Scientific Films
(Ian West)

167 Ed Jarzembowski

169 FLPA (SA Team/Fot Natura)

171 *left* FLPA (Mitsuhiko Imamori)

171 *right* FLPA (Mark Moffett)

173 *above* Premaphotos
(Ken Preston–Mafham)

173 *below* Premaphotos
(Ken Preston–Mafham)

175 Natural Visions

177 FLPA (Mark Moffett)

178 FLPA (Mark Moffett)

180 FLPA (Mark Moffett)

181 Nature Picture Library
(Premaphotos)

182 Michael Chinery

183 *above* Nature Picture Library
(Duncan McEwen)

183 *below* Nature Picture Library
(Andrew Harrington)

185 Michael Chinery

186 Nature Picture Library
(Ken Preston–Mafham)

187 Nature Picture Library
(Duncan McEwen)

189 Oxford Scientific Films
(Richard Packwood)

191 Oxford Scientific Films
(Michael Fogden)

193 *above* Oxford Scientific Films

193 *below* Michael Chinery

194 Auscape (Wayne Lawler)

195 *above* Oxford Scientific Films
(Kjell Sandved)

195 *below* Oxford Scientific Films

196 *above* Premaphotos
(Ken Preston–Mafham)

196 *below* Michael Chinery

198 *above* Premaphotos
(Ken Preston–Mafham)

198 *below* Premaphotos
(Ken Preston–Mafham)

199 Nature Picture Library
(Ken Preston–Mafham)

200 *above* FLPA (Jan van der Knokke)

200 *below* Oxford Scientific Films
(London Scientific Films)

203 Oxford Scientific Films
     (Paulo di Oliveira)

205 FLPA (Roger Tidman)

207 Nature Picture Library (Ken
     Preston-Mafham)

209 *above* Oxford Scientific Films
     (David Fox)

209 *below* Oxford Scientific Films
     (David Fox)

211 *above* Keith Brust

211 *below* Keith Brust

213 Natural Visions (Jeremy Thomas)

214 Nature Picture Library
     (David Tipling)

218 Gerald Cubitt

220 Premaphotos (Ken Preston-Mafham)

222 Oxford Scientific Films (Alan Root)

223 Nature Picture Library (Pete Oxford)

224 Nature Picture Library
     (Georgette Douwma)

226 Premaphotos (Ken Preston-Mafham)

228-9 Auscape (D & F. Parer-Cooke)

231 FLPA (Mitsuhiko Imamori)

233 Premaphotos (Rod Preston-Mafham)

234 Premaphotos (Ken Preston-Mafham)

235 Oxford Scientific Films
     (Philip J. Devries)

237 *above* Oxford Scientific Films
     (Michael Fogden)

237 *below* Oxford Scientific Films
     (Michael Fogden)

241 Nature Picture Library
     (Ken Preston-Mafham)

242 FLPA (Pete Oxford)

243 Oxford Scientific Films
     (Satoshi Kuribayashi)

244 Natural Sciences Image Library
     (Peter E. Smith)

249 *above* Nature Picture Library
     (Jim Clare)

249 *below* Simon Williams

250 FLPA (Hugh Clark)

255 FLPA (Mark Moffett)

257 Oxford Scientific Films

259 Nature Picture Library
     (David Tipling)

260 FLPA (Konrad Wothe)

261 Nature Picture Library
     (John Downer)

263 Oxford Scientific Films
     (Kathie Atkinson)

265 Oxford Scientific Films

266 Nature Picture Library
     (Ken Preston-Mafham)

267 Nature Picture Library
     (Martin Dohm)

268 FLPA (Michael & Patricia Fogden)

270 FLPA (Mitsuhiko Imamori)

271 Premaphotos (Ken Preston-Mafham)

273 FLPA (Mark Moffett)

274 FLPA (Mark Moffett)

275 FLPA (Mark Moffett)

Index

The names here – English or scientific – are those used in the text. Numerals in **bold** type indicate illustrations.